WHALING
on the NORTH CAROLINA COAST

MARCUS B. SIMPSON, JR., AND SALLIE W. SIMPSON

Raleigh

Division of Archives and History
North Carolina Department of Cultural Resources

1990

FOREWORD

Over the years the Historical Publications Section has published a number of titles related to the history of coastal North Carolina. Such booklets as *The Pirates of Colonial North Carolina* (16 printings) by Hugh F. Rankin, *North Carolina Lighthouses* (6 printings) by David Stick, and *The "Unpainted Aristocracy": The Beach Cottages of Old Nags Head* (4 printings) by Catherine W. Bishir are among the most popular titles ever published by the Division of Archives and History. *Whaling on the North Carolina Coast* by Marcus B. Simpson, Jr., and Sallie W. Simpson promises to join that list of best sellers.

Originally published as an article titled *The Pursuit of Leviathan: A History of Whaling on the North Carolina Coast* in the January, 1988, issue of the *North Carolina Historical Review*, this history of whaling provides a fascinating account of an economic enterprise rich in the romance of the sea. Whale fishing began as an important economic activity in the seventeenth century. In North Carolina, shore whaling prevailed, although during much of the nineteenth century the so-called Hatteras grounds attracted numerous pelagic whalers from New England. By the First World War, shore whaling in North Carolina had largely disappeared with the depletion of whales from overfishing. Still, the legends and lore persist, and the Simpsons have captured the sense of high adventure along the Outer Banks each time the cry "Thar she blows" went up.

Marcus Simpson, a native of Sanford, North Carolina, earned a B.S. degree at Davidson College and the M.D. degree from the University of North Carolina at Chapel Hill. He is associate professor of pathology at George Washington University. Sallie Simpson, a native of Washington, D.C., holds a B.A. degree from George Washington University and a B.S. degree from the University of Maryland, Baltimore County, where she is a graduate student in behavioral medicine and a candidate for the Ph.D. degree. Their article received the prestigious 1988 Robert D. W. Connor Award from the Historical Society of North Carolina for the best article published in the *North Carolina Historical Review*. Readers of this pamphlet will understand why.

Jeffrey J. Crow
Historical Publications Administrator

February, 1990

Whales have been a source of special fascination for man since pre-historic times. The largest creatures to inhabit the earth, these leviathans have inspired a sense of mystery and wonder that has often found expression in religion, literature, and works of art. From the biblical account of Jonah to Herman Melville's *Moby Dick* and from primitive cave paintings to contemporary sculpture, whales have captured man's creative attention as symbols of good and evil, of power and freedom, and, most recently, of the need for preservation at a global level. In recent years the threatened extinction of several whale species has prompted a renewed interest in the history of man's economic exploitation of these sea mammals, whose worldwide populations have, in some cases, been decimated by centuries of commercial whaling.[1]

Although whales have been hunted since antiquity, organized whaling ventures first began on a large scale in the early seventeenth century, when Dutch and British fleets actively promoted the industry at the Island of Spitzbergen in the Arctic Ocean.[2] A primary motivation for the British colonization of North America was the development of all forms of "fishing," and by the early 1700s New England had become established as the hub of American whaling. The large scope of whaling from that area has obscured the fact that some species of whales were actively hunted by local fishing communities along much of the Atlantic coast

[1]Ivan T. Sanderson, *Follow the Whale* (London: Cassell and Company, 1956), 17-18, hereinafter cited as Sanderson, *Follow the Whale*; M. E. Q. Pilson and E. Goldstein, "Marine Mammals," in *Coastal and Offshore Environmental Inventory: Cape Hatteras to Nantucket Shoals* (Kingston, Rhode Island: University of Rhode Island, Marine Publication Series 2, 1973), 7/1-7/48, hereinafter cited as Pilson and Goldstein, "Marine Mammals"; David K. Caldwell and Frank B. Golley, "Marine Mammals from the Coast of Georgia to Cape Hatteras," *Journal of the Elisha Mitchell Scientific Society*, 81 (May, 1965), 24-32, hereinafter cited as Caldwell and Golley, "Marine Mammals."

[2]F. C. Sanford, "Notes upon the History of the American Whale Fishery," in U.S. Congress, Senate, *United States Commission of Fish and Fisheries: Report of the Commissioner for 1882*, Forty-seventh Congress, Second Session, 1882-1883, Senate Miscellaneous Document No. 46, p. 205, hereinafter cited as Sanford, "American Whale Fishery"; Sanderson, *Follow the Whale*, 370-372; J. T. Jenkins, *A History of the Whale Fisheries* (London: H. F. and G. Witherby, 1921); Harry Morton, *The Whale's Wake* (Honolulu: University of Hawaii Press, 1982); Obed Macy, *The History of Nantucket* (Nantucket: Obed Macy, 1880), hereinafter cited as Macy, *History of Nantucket*.

from Maine to South Carolina during the colonial period.[3] In some localities, such as Williamsburg, Virginia, the pursuit of the great whales was apparently a rather casual and short-lived experiment.[4] In other regions, such as along the New Jersey shore, whaling ventures may have been a force promoting immigration and settlement.[5]

By the early nineteenth century, however, shore-based whaling had been largely abandoned along the Atlantic coast south of New York, with the notable exception of North Carolina. Whale hunting continued locally along the North Carolina coast until the twentieth century, in part because of the geographical position of the outer barrier islands. The close proximity of these sand banks to the Gulf Stream and to the migratory routes of certain whales provided a choice opportunity for whaling. North Carolina's local whaling industry centered around Cape Lookout, where, in their spring migration, northbound right whales passed close by the islands of Bogue and Shackleford Banks.[6]

Commercial American whaling actually began as a local shore-based activity during the seventeenth century, gradually evolving into an industry of open-sea fleets that hunted their quarry to the remotest oceans of the globe.[7] In contrast to their counterparts in New England, however,

---

[3]Sanderson, *Follow the Whale*, 182-217; A. Howard Clark, "The Whale Fishery," in George Brown Goode (ed.), *The Fisheries and Fishery Industries of the United States* (Washington: Commission of Fish and Fisheries, 5 sections, 1884-1887), Section V, Volume II, 26-65, 102-144, hereinafter cited as Clark, "Whale Fishery"; Alexander Starbuck, "History of the American Whale Fishery from its Earliest Inception to the Year 1876," in *Report of the Commissioner of Fish and Fisheries for 1875-1876* (Washington: Commission of Fish and Fisheries, 1878), Appendix A, 1-77, hereinafter cited as Starbuck, "American Whale Fishery."

[4]Some of the wealthy citizens of Virginia subscribed to outfit the sloop *Experiment* to engage in whaling in the spring of 1751. The vessel returned in May of that year with a recently captured whale. *Virginia Gazette* (Williamsburg), May 9, 1751; Starbuck, "American Whale Fishery," 43; Clark, "Whale Fishery," 107.

[5]George F. Boyer, *Cape May County Story Book I* (Egg Harbor City, New Jersey: Laureate Press, 1975), 7-11; Harold F. Wilson, *The Jersey Shore* (New York: Lewis Historical Publishing Co., 1953), 157-166; Paul Sturtevant Howe, *Mayflower Pilgrim Descendants in Cape May County, New Jersey* (Cape May: Albert R. Hand, 1921), 13; Harry B. Weiss, *Whaling in New Jersey* (Trenton: New Jersey Agricultural Society, 1974), hereinafter cited as Weiss, *Whaling in New Jersey*.

[6]The right whale (*Eubalaena glacialis*) was so named because it was the "right" species for the whale fishers. Right whales were relatively docile, slow moving, large animals that were found close to shore and remained afloat after being killed, making them good targets for the shore whalers in the early years of the industry. David Stick, *The Outer Banks of North Carolina* (Chapel Hill: University of North Carolina Press, 1958), 184-194, hereinafter cited as Stick, *Outer Banks*; Eugene P. Odus (ed.), *A North Carolina Naturalist, H. H. Brimley: Selections from His Writings* (Chapel Hill: University of North Carolina Press, 1949), 97-115, hereinafter cited as Odum, *H. H. Brimley Writings*; Gary S. Dunbar, *Historical Geography of the North Carolina Outer Banks* (Baton Rouge: Louisiana State University Press, 1958), 76, hereinafter cited as Dunbar, *North Carolina Outer Banks*; R. Edward Earll, "North Carolina and its Fisheries," in George Brown Goode (ed.), *The Fisheries and Fishery Industries of the United States* (Washington: Commission of Fish and Fisheries, 5 sections, 1884-1887), Section II, 490-491, hereinafter cited as Earll, "North Carolina Fisheries"; *News and Observer* (Raleigh), April 5, 1931; August 31, 1969, hereinafter cited as *News and Observer*; F. Ross Holland, Jr., *A Survey History of Cape Lookout National Seashore* (Washington: Department of the Interior, 1968), 11-19, hereinafter cited as Holland, *Survey History of Cape Lookout*; Pilson and Goldstein, "Marine Mammals," 10-12; Caldwell and Golley, "Marine Mammals," 29.

[7]Starbuck, "American Whale Fishery," 4-36; Clark, "Whale Fishery," 26-40; Frances Diane Robotti, *Whaling and Old Salem* (Salem, Massachusetts: Newcomb and Gauss Co., 1950),

The whaling industry began in North Carolina during the second half of the seventeenth century. Although by the early 1700s New England had become the hub of American whaling, whalers continued to hunt off the North Carolina coast for two and a half centuries. In 1725 the governor of North Carolina granted to one Samuel Chadwick of Carteret Precinct this license "to fish for Whale or other Royall fish." In exchange for the license Chadwick agreed to pay the governor 10 percent of the "oyle and bone." Photograph of the license from the files of the Division of Archives and History.

the whalers of North Carolina never made the transition to a pelagic version of the industry but continued instead to engage in shore whaling until the demise of the activity around World War I. Largely because of the early decline of shore whaling in New England, fishermen from Long Island and Massachusetts began to outfit seafaring voyages to more favorable locales, a pattern that foreshadowed their rise to international domination of whaling in the mid-nineteenth century. Compared to North Carolina, the New England area was in a stronger position to make such a transition, because of superior resources in shipbuilding, capital, markets, deepwater harbors, and sailing manpower. Thus, from similar initial methods of whale fishing during the seventeenth century, New England and North Carolina had, by the early eighteenth century, diverged into two fundamentally different approaches to whaling.[8]

Although never more than a minor industry in North Carolina, whaling was pursued along the coast for more than two and a half centuries. Throughout most of this time, whale hunting in Carolina waters followed

hereinafter cited as Robotti, *Whaling and Old Salem*; George Francis Dow, *Whale Ships and Whaling* (Salem, Massachusetts: Marine Research Society, 1925), hereinafter cited as Dow, *Whale Ships and Whaling*.

[8]Shore whaling also persisted along the northeastern coast until the twentieth century, with an active industry, for example, on eastern Long Island, New York. Randall R. Reeves and Edward Mitchell, "The Long Island, New York, Right Whale Fishery: 1650-1924," *Report of the International Whaling Commission*, Special Issue No. 10 (1987), 201-220, hereinafter cited as Reeves and Mitchell, "Long Island Right Whale Fishery"; E. J. Edwards and J. E. Rattray, *"Whale Off!": The Story of American Shore Whaling* (New York: Frederick A. Stokes, 1932); Starbuck, "American Whale Fishery," 4-77.

two distinct methods: shore-based whaling mostly by local residents using small rowing boats, and open-sea or pelagic whaling, conducted mainly by sailing ships from New England and New York. The local shore fishery initially relied on "drift" whales that washed ashore or whales that became stranded along the banks. That rather passive approach was supplemented and then largely replaced by the use of crews in open double-ended rowing boats to chase the whales, which were harpooned and towed to shore, where their blubber oil and whalebone were procured on the open beach. Such activities were conducted from late December to early June, although the peak season was usually from February to early May, during the northward spring migration of the right whale, the principal victim of the industry.[9]

In contrast, the offshore and pelagic whaling activities involved larger ocean-going vessels, mostly from northern ports in New York and Massachusetts, that visited the area at different times of the year depending on their quarry and itinerary. North Carolina's shore whaling activities were centered around Beaufort, with most crews active off Cape Lookout and Shackleford Banks. Although some northern offshore whaling vessels also hunted in the area southeast of Beaufort, they concentrated their major whaling activities during much of the nineteenth and twentieth centuries in the so-called "Hatteras ground." The ground lay between 35° to 38° north latitude and 70° to 75° west longitude along the edge of the Gulf Stream some distance off the northeast portion of the Carolina coast. The Hatteras ground provided opportunity for capturing sperm whales and perhaps an occasional right whale.[10]

Whalers killed both sperm and right whales for the oil from their blubber, which was used as a lubricant and fuel. Right whales also yielded up "whalebone," the flexible but tough baleen used by the whale to filter its food from seawater. Whalebone was sold commercially for products ranging from buggy whips to corset stays. Sperm whales' heads contain a unique "spermaceti" oil, which was of a superior quality for various uses, including eventually the manufacture of premium candles. Once the oil, tongue, baleen, and external layer of blubber were removed,

---

[9]Stick, *Outer Banks*, 185, 190; Dunbar, *North Carolina Outer Banks*, 127; Earll, "North Carolina Fisheries," 490; Holland, *Survey History of Cape Lookout*, 11-17; Odum, *H. H. Brimley Writings*, 109.

[10]The sperm whale (*Physeter macrocephalus*) was the species that formed the basis for most of the eighteenth-century growth in the American whale fishery. Other species reportedly taken commercially in North Carolina were the fin whale (*Balaenoptera physalus*) and the humpback whale (*Megaptera novaeangliae*), although no specific records of captures appear to have been documented. Clark asserted that the right whales were also found in the Hatteras ground, but again there is apparently no specific documentation of this claim. Pilson and Goldstein, "Marine Mammals," 15-18, 20-22, 35-37; Clark, "Whale Fishery," 8-9, 15-16, 22, 24, 48-49, 65, 144-145; George Brown Goode, "The Whales and Porpoises," in George Brown Goode (ed.), *The Fisheries and Fishery Industries of the United States* (Washington: Commission of Fish and Fisheries, 5 sections, 1884-1887), Section I, 8, hereinafter cited as Goode, "Whales and Porpoises"; Randall R. Reeves and Edward Mitchell, "American Pelagic Whaling for Right Whales in the North Atlantic," *Report of the International Whaling Commission*, Special Issue No. 10 (1987), 233-234, hereinafter cited as Reeves and Mitchell, "American Pelagic Whaling for Right Whales."

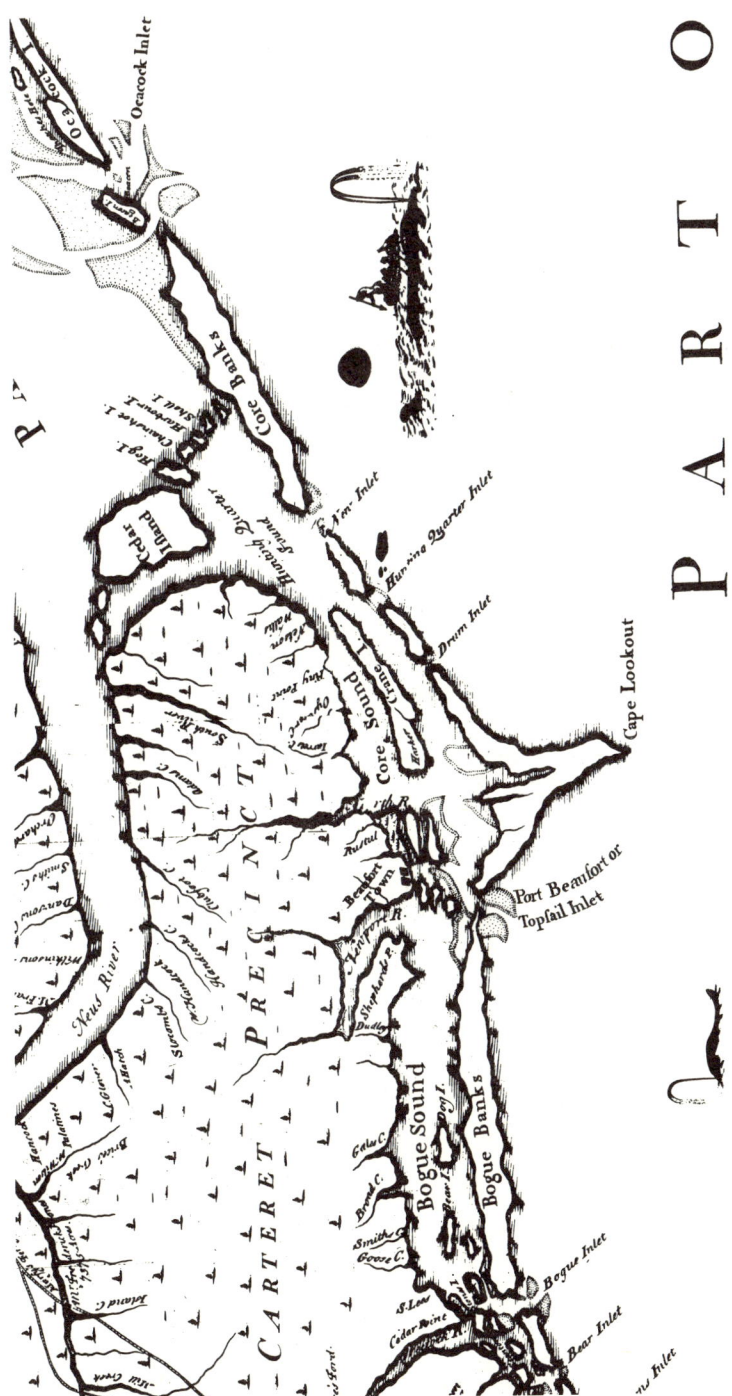

P A R T O

Whaling in North Carolina centered on the region between Ocracoke Inlet and Bogue Inlet, with most crews active off Cape Lookout and Shackleford Banks. This portion of a 1733 map by Edward Moseley, surveyor general of the colony, shows the area between the two inlets. Drawings of whales and whalers embellish the map, which is reproduced here from W. P. Cumming, *North Carolina in Maps* (Raleigh: State Department of Archives and History, 1966), plate VI.

the bulk of the whale—consisting of muscle, internal organs, and bones—was usually discarded.[11]

Details concerning the inception of whaling in North Carolina remain lost in obscurity, but whales along the coast were considered worthwhile commercial objects by the second half of the seventeenth century. Among the earliest documents indicating commercial whaling was a license granted in 1666 by Peter Carteret—assistant governor of the North Carolina colony, then called Albemarle—to allow three New England men to take whales along the northeast coast of North Carolina. Carteret's permit was recorded in 1667 at Southampton, Long Island, and gave to Humphrey Huse (Hughes) and John Cooper of Southampton and Nicholas Stevens of Boston the following rights: "to enjoy the privilege to make use of all the whales that shall be cast up or that they can use anyways to kill or destroy, between the inlet of Roanoak and the inlet of Caretuck." The group intended to sail in the sloop *Speedwell*, from Southampton, New York, to search for whales in Carolina waters, but whether the men took advantage of their North Carolina license is unclear. Hughes later settled on the New Jersey coast.[12] Other northeastern whalers were also interested in the southern coast during that time, for in August, 1688, one Timotheus Vandernen, commander of the brigantine *Happy Return* of New York, petitioned for permission to outfit and sail for sperm whales in the Bahamas and along the coast of Florida. That course would probably have required his transit through North Carolina waters.[13]

The extent of such early northeastern whaling ventures along the Carolina shore cannot be accurately defined. Local shore-based whale processing, however, was well established by the 1660s and 1670s, for among the major products of the Albemarle colony were whale oil and whalebone, as indicated by correspondence and shipping account records. These reports are not detailed enough to reveal whether the whales were actively pursued and killed from boats or simply obtained as dead or stranded animals. In 1668 Colleton Island produced about eighty barrels of whale "Oile" and on October 22, 1668, Peter Colleton, who two years earlier had become one of the proprietors of the colony, wrote to Peter Carteret: "I am glad you have found out a Whale ffishing I doe desire you to Continue it ffor our Shipps yt went to Greeneland tooke but one Whale last season soe that Oyle & Whalebone is at psent a great Comodity here & I conceive London is a better markett than Barbados, soe yt I desire you to send all you take here, And consigne it to mee, all wch I shall Lay out in supplyese for your plantacon & return to you." To the proprietors in 1672-1673, Peter Carteret reported in abstracts of goods from plantations at "Colliton Island" and "Powells Point" that "Att severall tymes I have Shipped one

---

[11]Starbuck, "American Whale Fishery"; Clark, "Whale Fishery."

[12]Stanley Williamson, "Origin of the Hughes Family in Cape May County, N. J.—and the Seven Humphreys," *Cape May County Magazine of History and Genealogy*, II (June, 1942), 144-145; Weiss, *Cape May Whaling*, 21; Reeves and Mitchell, "American Pelagic Whaling for Right Whales," 228.

[13]Starbuck, "American Whale Fishery," 15.

hundred nynty five Barrells of Whale Oyle for London & Consigned it to Sir Peter Colliton for yor Lordship & Company, wch I Conceive may have Cleered about 25. s p barrel" or a total of over 243 pounds sterling value. In 1673 London merchant John Buckworth notified the proprietors that Peter Carteret had shipped a total value of 241 pounds sterling in "whale Oyle" to Sir Peter Colleton for "yor Lordshipps accompt."[14]

This rising economic importance of whale products in North Carolina naturally prompted a need for regulation, taxation, and proprietary interest in encouraging the industry. The "Fundamental Constitutions of Carolina," drafted by orders of the proprietors in 1669, included the notation that all whale fishing and half of all ambergris, a valuable product of sperm whales, "shall wholly belong to the Lords Proprietors."[15] The proprietors soon chose, however, to promote whale fishing in North ᑕarolina by forgoing their rights to these products, as on July 13, 1681, when they instructed Governor John Jenkins and the executive council to make public an announcement that the inhabitants of the colony were authorized to take "what whales they can and convert them to their owne use" during the next seven years. The proprietors had been "informed that there are many Whales upon the Coast of Carolina" and thus wished to allow the "incouragement of Carolina."[16] Subsequently, in 1691, the proprietors extended for another twenty years their previous permission that Carolinians could make unrestricted use of whales, the Fundamental Constitutions notwithstanding.[17]

In the fifteen-year period after that extension of whaling rights by the proprietors, evidence of an active whale industry in North Carolina appeared in court records of complaints that quite predictably arose when economic matters were at stake. Such disputes over the ownership of whales and whale products were common throughout the colonies during that time, as in New York and Massachusetts, where lawsuits also provide valuable documentation of local whaling endeavors.[18] Court

[14]Mattie Erma Edwards Parker (ed.), *North Carolina Higher-Court Records, 1670-1696*, Volume II of *Colonial Records of North Carolina, Second Series*, edited by Mattie Erma Edwards Parker and others (Raleigh: Division of Archives and History, Department of Cultural Resources, projected multivolume series, 1963—), xx, hereinafter cited as Parker, *North Carolina Higher-Court Records*, II; William S. Powell (ed.), *Ye Countie of Albemarle in Carolina* (Raleigh: State Department of Archives and History, 1958), 31-32, 57, 59, 62, hereinafter cited as Powell, *Ye Countie of Albemarle*.

[15]William L. Saunders (ed.), *The Colonial Records of North Carolina* (Raleigh: State of North Carolina, 10 volumes, 1886-1890), I, 205, hereinafter cited as Saunders, *Colonial Records*.

[16]Powell, *Ye Countie of Albemarle*, 31; Saunders, *Colonial Records*, I, 338; Robert J. Cain (ed.), *Records of the Executive Council, 1664-1734*, Volume VII of *Colonial Records of North Carolina, Second Series*, edited by Mattie Erma Edwards Parker and others (Raleigh: Division of Archives and History, Department of Cultural Resources, projected multivolume series, 1963—), 359-360, hereinafter cited as Cain, *Records of the Executive Council*, VII.

[17]Robert J. Cain (ed.), *North Carolina Higher-Court Minutes, 1724-1730*, Volume VI of *Colonial Records of North Carolina, Second Series*, edited by Mattie Erma Edwards Parker and others (Raleigh: Division of Archives and History, Department of Cultural Resources, projected multivolume series, 1963—), xxi, hereinafter cited as Cain, *Higher-Court Minutes*, VI.

[18]Starbuck, "American Whale Fishery," 10, 34.

records from the 1690s suggest that the Carolina shore fishery had become rather competitive, for a number of litigants were embroiled in charges and countercharges involving whales. These reports do not indicate whether the whales were obtained by active pursuit and harpooning or merely by processing drift whales.

In 1694 a certain Charles Thomas asked the court to order one Mathias Towler to reimburse him for labor costs of two shillings per day incurred when "your petitioner Workeing upon A Whale in March last ten dayes and att last Mathias Towler Tooke the said Whale from your petitioner and never made any Satisfaction to your petitioner."[19] Mathias Towler complained to Governor Philip Ludwell, however, that he had received a previous court order instructing Charles Thomas and one Timothy Pead to return a whale that they had wrongly taken from him, noting that Thomas and Pead had "by Forse and violence driven your Petitioner from the Whale and Contemtiously detaines and keep away your petitioner from the Whale." Towler had been given a "Lycence from the Honorable Governor" for whaling,[20] but the court held that Pead and Thomas had committed no wrong and in fact agreed that "the said Charles Thomas had done ten dayes Worke upon the Whale which afterwards Mathias Towler Tooke from Timothy Pead and Company." The court ordered Towler to pay Thomas and Pead for their work on the whale and, furthermore, instructed Towler to compensate one Anne Ros for her labor, for she and her family had "tryed up three Barrell of Oyle out of the whale which Mathias Towler afterwards took from Timothy Pead and their Company."[21] Towler continued to have legal problems relating to his whaling, as in 1697 when he was judged to be indebted to Colonel Thomas Pollock for three barrels of "Whale Oyle," which was to be delivered to a "convenient Landing near his the said Toulers house at the sand Banckes."[22]

Other court records indicate that whale products had begun to be used as a form of payment in North Carolina during the proprietary period. In his 1697 suit against a certain James Carroon, one Thomas Harvey asked that the debt be paid in "whale Oyle"; and Harvey further requested that ten shillings owed him by one William Steel likewise be paid in "oyl." Steel's thirty-shilling debt to his creditor Thomas Durant was also to be settled using "Oyle" as currency.[23] Whale products were shipped out of the colony and apparently accepted as exchange currency. The sloop *Peter*, for example, had whale oil as a cargo item upon departure in

---

[19]Parker, *North Carolina Higher-Court Records*, II, 74.

[20]Parker, *North Carolina Higher-Court Records*, II, 18-20.

[21]Parker, *North Carolina Higher-Court Records*, II, 42-44; Saunders, *Colonial Records*, I, 419.

[22]Mattie Erma Edwards Parker (ed.), *North Carolina Higher-Court Records, 1697-1701*, Volume III of *Colonial Records of North Carolina, Second Series*, edited by Mattie Erma Edwards Parker and others (Raleigh: Division of Archives and History, Department of Cultural Resources, projected multivolume series, 1963—), 414-415, hereinafter cited as Parker, *North Carolina Higher-Court Records*, III.

[23]Parker, *North Carolina Higher-Court Records*, III, 122, 204-205, 207.

1697,[24] and in 1705 Robert Quary and the "Pensylvania Company" filed a complaint that John Neal's debt to them should be paid using whalebone or oil.[25] One Carolinian, Mary Guthrie, accepted whalebone in exchange for "Two foules" in 1713, and the assembly included whale oil among the commodities officially recognized as provincial currency in 1715 and 1719.[26]

A different perspective on whales and whaling was provided by John Lawson, surveyor general of North Carolina, whose observations from the early 1700s were conveyed in his *New Voyage to Carolina*. Lawson noted that "Whales are very numerous, on the Coast of North Carolina, from which they make Oil, Bone &c. to the great Advantage of those inhabiting the Sand-Banks, along the Ocean, where these Whales come ashore, none being struck or kill'd with a Harpoon in this place, as they are to the Northward, and elsewhere; all those Fish being found dead on the Shoar, most commonly by those that inhabit the Banks, and Sea-side, where they dwell, for that Intent, and for the Benefit of Wrecks." Lawson's careful observations strongly suggest that Carolina shore whaling remained largely and perhaps entirely a matter of processing stranded or drift whales until some time after the first decade of the eighteenth century. Lawson also reported that whalebone and oil were major products of the colony and that oil was being made not only from certain whales but also from "Porpoises," "Crampois," "Thrashers," and various species of fish.[27]

Lawson further stated that four different types of whales were to be found on the coast, but the "Sperma Cati" or sperm whale was the only species that he described sufficiently to permit confident identification. Lawson asserted that the sperm whale was the choicest and richest of the whales. Unfortunately, they were rarely seen in Carolina, for he "never heard of but one found on this Coast, which was near Currituck-Inlet." Lawson mentioned another sort of whale that had likewise only been seen once on the Carolina coast, it being "contrary, in Shape, to all others ever found before" and measuring sixty feet in length but only three or four feet in diameter. From the other two species, which Lawson called the "Bottle-Nosed" and the "Shovel-Nose" whales, local fishermen obtained whalebone and oil.[28] One of those species may have been the right whale,

[24]Parker, *North Carolina Higher-Court Records*, III, xix; Miscellaneous Papers, Colonial Court Records, 192, Archives, Division of Archives and History, Raleigh, hereinafter cited as CCR with appropriate number.

[25]William S. Price, Jr. (ed.), *North Carolina Higher-Court Records, 1702-1708*, Volume IV of *Colonial Records of North Carolina, Second Series*, edited by Mattie Erma Edwards Parker and others (Raleigh: Division of Archives and History, Department of Cultural Resources, projected multivolume series, 1963–), 162, hereinafter cited as Price, *North Carolina Higher-Court Records*, IV.

[26]Cain, *Records of the Executive Council*, VII, 471; Cain, *Higher-Court Minutes*, VI, xxii.

[27]John Lawson, *A New Voyage to Carolina*, edited by Hugh Talmage Lefler (Chapel Hill: University of North Carolina Press, 1967), 157-158, 166, hereinafter cited as Lawson, *New Voyage to Carolina*.

[28]Lawson, *New Voyage to Carolina*, 157-158.

but Lawson's account is sufficiently confusing that the species of the two whales cannot be confidently established.

In the two decades following Lawson's report, whaling activities assumed an increasing importance in North Carolina. Although the proprietors presumably wished to benefit from the profits of whale fishing, they decided to forgo some immediate return to promote long-term growth of the industry. Proprietary instructions to governors Edward Hyde in 1712 and Charles Eden in 1713 included the reminder that Parliament had lowered the duties on whale products to encourage the fish industry.[29] In 1712, however, the proprietors instructed Daniell Richardson, receiver general of North Carolina, that he was to "take into . . . possession our Share of . . . Ambergrease" and other such items that "of right belong to us."[30]

By 1714 Governor Eden had responsibility for supervising the collection of the "Tenths" of whale oil and bone owed to the government. During much of Eden's tenure as governor from 1714 to 1722, one "Cap. John Ricords" or "Records" was among the most successful whalers in Carolina waters. Captain "Ricords, Cap. Thomas, & Others whaleing on our Sea coast" allegedly paid some £2,000 duty to the government, indicating that whale products amounting to at least £20,000 value had been taken and reported as required by law.[31] Other highly productive whalers included Captain Josiah Doty, equipped with several boats and a large crew, who captured many whales and extracted 300 barrels of oil and 1,000 weight of whalebone during the 1727 season.[32]

The proprietors also wanted to promote North Carolina whaling by encouraging men from other colonies to fish along the coast. Their inducements proved to be successful, and in the last fifteen years as a proprietary colony, North Carolina became the site of whale hunting by New Englanders. In 1715 one of the proprietors, John Carteret, wrote to Governor Eden instructing him to "give all due Encouragement to such persons as are willing to come and settle among you" by granting a "Power or Liberty to any New England Men or others to catch Whale . . . or any other Royal Fish upon your Coast, during the Term of three years, they paying only two Deer Skins yearly to the Lords as an acknowledgment" for the privilege.[33] This decision came at a propitious time, for New England shore whaling was beginning to decline and efforts there were being directed at outfitting seaworthy vessels of about thirty to thirty-six tons to chase whales in the deep ocean, particularly along the Gulf Stream to the southward. According to one apocryphal story, the capture of a sperm whale by Captain Christopher Hussey off the coast of Nan-

---

[29]Cain, *Records of the Executive Council*, VII, 449, 471.

[30]Cain, *Records of the Executive Council*, VII, 457.

[31]Petition of Richard Everard to Edmund Porter, July 23, 1730, Vice-Admiralty Papers, CCR 142.

[32]Jean Bruyere Kell and Thomas A. Williams (eds.), *North Carolina's Coastal Carteret County during the American Revolution* (Greenville: Era Press, 1975), 105, hereinafter cited as Kell and Williams, *Carteret County*.

[33]Saunders, *Colonial Records*, II, 175-176.

tucket in 1712 "gave new life to the business," and soon the New Englanders were engaged in cruises of about six weeks' duration in North Atlantic waters.[34] This combination of events was apparently a boon to Carolina whaling, as revealed for example by a three-week period in 1719 when four New England vessels brought more than 500 barrels of oil and considerable whalebone from North Carolina to Boston.[35]

In 1723 the proprietors instructed the council and assembly to take "Care of the Fishery" not only as a means of improving trade but in order to "cause a great Number of Inhabitants to come to the province and settle among you."[36] Such inducements may have contributed to the decision of four New England whalers to move to North Carolina. In 1725 Samuel Chadwick acquired 130 acres of land at Straits, North Carolina, where he identified himself as "late of New England but now an inhabitant of Cartaret Precinct." By 1726 Chadwick was accompanied by Ephraim Chadwick, Ebenezer Chadwick, and John Burnap, all formerly of New England but now residents of Carteret Precinct. Their intention in relocating to North Carolina was revealed in the license issued by Richard Rustull: "To Samuel Chadwick you are hereby permitted with three boats to fish for whale or other Royall fish on the Seay Coast by this Government and whatsoever ye shall catch to convert to your own use paying to ye Honble ye Governor one tenth parte of ye Oyle and bone made by Vertue of this License."[37] Just how long Chadwick and his associates continued·to engage in whaling is unclear, but they all eventually abandoned the activity for other pursuits.[38]

The North Carolina-New England connection was not always tranquil, however, and a number of controversies and problems occurred. In 1720, for example, Captain Samuel Butler sailed in a whaling sloop from New England to procure a license for whale fishing in North Carolina. Butler stopped at Hampton, Virginia, where Henry Irwin, a naval officer, told him that the governor of North Carolina had no power to grant whaling licenses in Carolina waters. Instead, Irwin allegedly informed Butler, the governor of Virginia had sole authority for whaling as far south as Cape Fear. Irwin threatened to seize any whale "Oyle or Fynn" that Butler took in North Carolina if he failed to obtain the correct license. Governor Eden and the council urged the proprietors to take action in England to prevent such detriments to the trade rightfully belonging to North Carolina.[39] Other problems with Virginia arose in 1722, when a sloop owned

---

[34]John R. Spears, *The Story of the New England Whalers* (New York: Macmillan Company, 1908), 67, hereinafter cited as Spears, *New England Whalers*; Starbuck, "American Whale Fishery," 20-23; Dow, *Whale Ships and Whaling*, 18-19; Sanderson, *Follow the Whale*, 212.

[35]Cain, *Higher-Court Minutes*, VI, xxii.

[36]Cain, *Records of the Executive Council*, VII, 505; Saunders, *Colonial Records*, II, 490.

[37]Samuel Chadwick was believed to have come from Falmouth, Massachusetts, perhaps by way of Nantucket. Amy Muse, *Grandpa Was a Whaler* (New Bern: Owen G. Dunn Co., 1961), 3-8, hereinafter cited as Muse, *Grandpa Was a Whaler*; *North Carolina Historical and Genealogical Register*, II (April, 1901), 298.

[38]Muse, *Grandpa Was a Whaler*, 12-21, 24-25, 53.

[39]Cain, *Records of the Executive Council*, VII, 103; Saunders, *Colonial Records*, II, 397.

by Thomas Brown of Virginia was seized by the North Carolina Court of Vice-Admiralty in Edenton for possession of counterfeit and forged clearance documents involving barrels of whale "oyl" and bone.[40]

On other occasions, visiting fishermen reaped the harvest of whales along the Carolina coast but departed without bothering to pay their due tax to the colony. In the 1730s John Brickell recounted one such incident. "Not many years ago," Brickell wrote, two northern whalers came to Ocracoke and took back some 340 barrels of whale oil and considerable whalebone. Unfortunately, "these Fishermen going away without paying the Tenths to the Governor, they never appeared to fish on these Coasts afterwards." Brickell noted that the "middling Whales . . . commonly yield seventy, eighty, or ninety Barrells of Fat or Oil"; and he further observed that "the People in these parts are not very much given to Industry, but wait upon Providence to throw those dead Monsters on Shoar, which frequently happens to great advantage and profit."[41]

A different sort of problem with New England trade occurred in 1734, when the sloop *Middleborough*, Edward Fuller, master, was impounded at Port Beaufort because of improper customs house certificates from its origin at Boston. Among the items on board were "4 Lances for Whaleing, 2 Spades for blubber, 39 empty barrels," and "2 hump spears." Although this equipment was perhaps being carried to North Carolina for sale to local whalers, it is also possible that Fuller intended to engage in whaling during the sojourn.[42]

Whaling activities also became a matter of public record during the stormy period of North Carolina's transition from a proprietary to a royal colony. At that time, most of the information concerning the whale fishery was associated with the acrimonious and often bitter conflict between governors George Burrington and Richard Everard during the late 1720s and early 1730s.[43] When appointed governor in 1723, Burrington was granted a lease for the whale fishery by the proprietors, who issued to Burrington, Chief Justice Christopher Gale, and John Lovick, secretary of the province, the "power and Liberty of Fishing and takeing all sorts of Whales" and "other Royall Fish whatsoever upon the Coast of the Northern part of the Province of Carolina."[44]

Burrington's strong-willed and often violent style contributed to his replacement by Richard Everard in 1725. Burrington maneuvered to be reinstated and was subsequently appointed as the first royal governor in

---

[40]Petition of Richard Everard to Edmund Porter, July 23, 1730, Vice-Admiralty Papers, CCR 142.

[41]John Brickell, *The Natural History of North-Carolina* (Dublin: James Carson, 1737; Murfreesboro: Johnson Publishing Company, 1968), 218-220, hereinafter cited as Brickell, *Natural History*.

[42]Inventory of the Goods Belonging to the Sloop Called the Middleborough, February 15, 1735; The Value of the Sloop Middleborough and All Her Appurtenances . . . , March 27, 1735, Vice-Admiralty Papers, CCR 142.

[43]Carl W. Ubbelohde, Jr., "The Vice-Admiralty Court of Royal North Carolina, 1729-1759," *North Carolina Historical Review*, XXXI (October, 1954), 517-528, hereinafter cited as Ubbelohde, "Vice-Admiralty Court"; Cain, *Higher-Court Minutes*, VI, lvi.

[44]Cain, *Records of the Executive Council*, VII, 502-504.

January, 1730. Burrington did not arrive in North Carolina until February, 1731, however, and his absence through 1730 gave Everard the chance to attack Burrington's supporters. Much of this fight occurred in the Court of Vice-Admiralty, where Everard filed charges that alleged violations of the laws pertaining to whaling. Many, but not all, of these charges were directed against Burrington's colleagues and associates.[45]

In July, 1730, Everard informed Edmund Porter, judge of the Court of Vice-Admiralty, that when Charles Eden became governor and vice-admiral in 1714, he began to receive the tenths of whale oil and bone due as tax "to the value of two thousand Pounds from Cap. John Ricords, Cap. Thomas, & Others whaleing on our Sea coast." Everard charged that Eden gave these proceeds to John Lovick, who should be required to pay the past-due amount.[46]

Everard also filed a complaint that William Reed, while governor, had for a period "upwards of two years" assumed the "right and power of receiving the Tenths of Whale Oyl & Bone and accordingly did demand and collect the same from Capt John Records & others whaleing on the Sea Coast of Port Beaufort to the Quantity of Sixty Barrels of Train Oyl and Eight hundred wt. of Bone being the Tenths due to his Majesty which he the sd Willm Reed converted & applied to his own use to the time of his death." Everard valued the amount past due at £300 and demanded that the account be paid from Reed's estate.[47]

Everard further charged that William Little, receiver general, had in January, 1729, empowered Ebenezer Harker of Port Beaufort to receive the tenths of whale oil and bone taken on the coast. Captain John Records and his company, apparently the major whaling group in the area during much of the preceding twenty years, had reportedly collected 670 barrels of oil and given sixty-seven to Little during the period in question, but Little had reported only eight barrels received, thereby defrauding the crown.[48]

In a seemingly unrelated case, Everard requested that the Court of Vice-Admiralty issue a warrant for the arrest of David O'Sheal, originally from Virginia but then practicing law in the North Carolina courts. David O'Sheal had induced Everard to appoint his brother Ben O'Sheal as naval officer in charge of Port Beaufort in November, 1728, in which capacity he was empowered to "receive the Tenths of Whale Oyl & Bone

---

[45]William S. Powell (ed.), *Dictionary of North Carolina Biography* (Chapel Hill: University of North Carolina Press, projected multivolume series, 1979—), I, 283-284; II, 171-172, hereinafter cited as Powell, *DNCB*; Ubbelohde, "Vice-Admiralty Court," 524-528.

[46]Petition of Richard Everard to Edmund Porter, July 23, 1730, Vice-Admiralty Papers, CCR 142.

[47]Petition of Richard Everard to Edmund Porter, July 23, 1730, Vice-Admiralty Papers, CCR 142.

[48]Petition of Richard Everard to Edmund Porter, July 23, 1730, Vice-Admiralty Papers, CCR 142. William Little responded by obtaining an order from William Smith, chief justice of North Carolina, prohibiting anyone, including the Vice-Admiralty Court, from further harassing him. The court then decided to drop the charges. Order of William Smith to Edmund Porter, April 5, 1731; Order of Court of Vice-Admiralty, April 6, 1731, Vice-Admiralty Papers, CCR 142.

Catched on the Sea Coast of the Said Port." During his appointment, Ben O'Sheal had reportedly collected eighty barrels of oil and half a ton of bone, a total value of £500 sterling; but after converting the duty to his own use he fled to Gambia. David O'Sheal was arrested and ordered to pay the amount owed by his brother, for whom he had earlier posted a bond.[49]

Everard may have had political motives for some of his actions; nevertheless, confusion and difficulties plagued the interpretation and proper enforcement of the whale "tax." In 1728, for example, the proprietors had to petition the king that payment of their tenth of the whale fishery was £800 sterling in arrears.[50] The council session in March, 1730, noted that "the Tenth of Whale Oyle and Bone taken on the Sea Coast of this Province which has been lately claim'd by the Jud[g]e of the Ad[m]iralty of this Province as a Droit of Ad[m]iralty," actually, according to law, belonged to the governor of the colony.[51] In the same year, Governor Burrington's commission as vice-admiral indicated his responsibility over all maritime matters, including royal fishes, whales, porpoises, dolphins, and any other fishes that have "in them a great or huge thickness or fatness."[52]

Also in 1730, the crown's instructions to Burrington noted that "for Some Years past, the Governors of Some of Our Plantations have Seized and appropriated . . . the produce of Whales . . . taken upon those Coasts, upon pretence that Whales are Royal Fishes, which tends greatly to Discourage this Branch of Fishery . . . and prevent persons from Settling there, It is therefore Our Will . . . that You do not pretend to any Such Claim, nor give any . . . Discouragment to the Fishery of Our Subjects. . . ."[53]

Whales continued to be a part of the battle fought between the Everard and Burrington factions in 1731 and 1732. Everard's attempt to use Judge Edmund Porter and the Court of Vice-Admiralty against Burrington's friends provided Burrington an additional incentive for attacking Porter. In May, 1731, the council at Edenton, with Burrington as governor, heard complaints against Porter for his part in the "notoriously wrong" and unjust suit brought by Everard against William Little, who had been charged improperly with failing to report the tenths of whale oil and bone that he had collected for the king.[54] Not intending to yield the issue, Everard complained to the king that "Mr. Burrington contrary to common Justice has . . . received and converted to his owne use the Arrears of perquisite due . . . from the Tenths of the Whale Fishery."[55] When Gabriel

---

[49]Petition of Richard Everard to Edmund Porter, July 23, 1730, Court Order for David Osheal, August 3, 1730, Vice-Admiralty Papers, CCR 191; Order of Edmund Porter to Marshal of Court, August 3, 1730, Vice-Admiralty Papers, CCR 142.

[50]Saunders, *Colonial Records*, II, 722.

[51]Cain, *Records of the Executive Council*, VII, 185-186; Saunders, *Colonial Records*, III, 213-214.

[52]Cain, *Records of the Executive Council*, VII, 582.

[53]Cain, *Records of the Executive Council*, VII, 597-598; Saunders, *Colonial Records*, III, 99.

[54]Cain, *Records of the Executive Council*, VII, 200; Saunders, *Colonial Records*, III, 231.

[55]Cain, *Records of the Executive Council*, VII, 256.

Johnston replaced Burrington as governor in 1734, the controversy finally ended, but government records of North Carolina whaling were seldom again so rich.[56]

The final whaling legacy of the Everard-Burrington era resides in the "New and Correct Map of the Province of North Carolina," produced in 1733 by Edward Moseley, who had been appointed to the North Carolina-Virginia boundary commission by Everard.[57] Moseley's map is embellished with an intriguing scene of a small rowboat crew in pursuit of a whale, the setting just off Core Banks. A second whale is shown near Cape Lookout. In 1737 Moseley's map and the two whales were plagiarized in the *Natural History of North-Carolina*, by John Brickell, sometime physician to Governor Everard.[58] Although the John White and Theodore de Bry map of 1590 had depicted whales or sea monsters along the North Carolina coast, those appear to have been merely decorative in purpose, in contrast to the whales on the Moseley map.[59]

These various records indicate that during the proprietary period whaling had become firmly established on the North Carolina coast, where the dual pattern had emerged of local shore-based whalers and northeastern pelagic whalers. Although the economic significance cannot now be clearly defined, North Carolina whaling had become lucrative enough to provoke lawsuits among the citizens and to tempt government officials into fraud and embezzlement of tax revenue. Whaling was sufficiently important to the proprietary colony that political factions seized the opportunity to exploit any controversy in the industry for their own gains through the judicial system. Whale oil had been granted recognition as an official medium of currency; and during the last fifteen years as a proprietary colony, North Carolina whalers may have obtained as much as 4,200 barrels of whale oil. The extent of whaling activities justified the attention of local naturalists and cartographers, as well as commentary from the New England industry.

Unfortunately, extrapolations cannot be confidently made about the number of whales taken during the proprietary period nor about the relative importance of drift whales, shore-based captures, and pelagic whaling. Little is known about the size and sex of the harvested whales, the efficiency of the "trying out" process, or variations in barrel size. Extant records mostly involve litigation and political squabbles, so that the combatants may have misrepresented the harvest size. The actual

---

[56]Gabriel Johnston (1699-1752), physician and scholar, served as governor from 1734 to 1752, the longest term of any North Carolina governor. Beth G. Crabtree, *North Carolina Governors, 1585-1974: Brief Sketches* (Raleigh: Division of Archives and History, Department of Cultural Resources, third edition, 1974), 34-35.

[57]Edward Moseley, "A New and Correct Map of the Province of North Carolina" (London: J. Cowley Sculp, 1733); Marcus B. Simpson, Jr., "Copperplate Illustrations in Dr. John Brickell's *Natural History of North-Carolina* (1737): Sources for the Provincial Map, Flora, and Fauna," *North Carolina Historical Review*, LXII (April, 1985), 123, hereinafter cited as Simpson, "Brickell's Sources."

[58]Simpson, "Brickell's Sources," 119-124.

[59]William P. Cumming, *The Southeast in Early Maps* (Chapel Hill: University of North Carolina Press, 1958), 101, and plates 51-54.

PURSUIT OF THE GREENLAND WHALE.

North Carolina's shore fishery initially relied on "drift" whales that washed ashore or became stranded. Eventually, however, whalers harpooned their prey in Carolina waters in the same manner shown in this engraving of a capture of a Greenland or Bowhead whale. For North Carolinians whaling was primarily a shore-based operation using double-ended row boats, but sailing vessels from New England and New York also engaged in open-sea or pelagic whaling off the Tar Heel coast. From *Harper's New Monthly Magazine*, XII (March, 1856), 477.

time interval of most reports is vaguely stated, and official export records are sparse. However, if one accepts the conventional assertion that an "average" right whale would yield 30 to 50 barrels of oil, then the North Carolina production of 4,200 barrels in 15 years would have represented from 84 to 140 whales, or 6 to 9 per year. Records of individual annual yield could run as high as 16 to 26 whales, assuming the 800 barrels reported during O'Sheal's tenure as collector were obtained from one season's whaling. Despite the inadequacies of such records, whaling had clearly become a steady enterprise on the North Carolina coast by the time the colony was transferred to the crown.[60]

In the remaining decades of the colonial period, whaling continued to grow in economic importance for the Northeast. Whales were becoming scarce in the waters off New England, and during the 1730s and 1740s Nantucket vessels began extending their activities ever more heavily to the southward, cruising the area in forty- to fifty-ton schooners until around the first of July each year.[61] By the 1750s, two innovations

---

[60]For a discussion of the complexity of extrapolating whale harvests based on oil production, see Reeves and Mitchell, "Long Island Right Whale Fishery," 208-209.

[61]Starbuck, "American Whale Fishery," 23.

contributed even greater impetus to such voyages, namely the installation of tryworks on board the ships and the development of methods for extracting spermaceti from sperm whale heads. The inclusion of tryworks on the whaling vessels permitted immediate boiling and extraction of the whale oil at the time of capture, where previously it was necessary to return to shore for trying out the blubber. This process greatly increased the carrying capacity of the ships and, more importantly, freed the whalers for voyages of years' duration rather than the previous maximum of a few months. The technology for large-scale extraction of spermaceti provided the basis for producing the highest quality candles available, and New England spermaceti candles were soon in heavy demand throughout the colonies and abroad.[62]

These incentives for catching sperm whales probably increased the whaling activities along the North Carolina coast, for in 1755 it was stated that sperm whales were "most plenty upon the coast of Virginia and Carolina."[63] The opportunities were not without risk, however, and ships plying the southeast coast fell victim to pirates and foul weather. In 1747 a whaling schooner from Boston and a sloop from Nantucket were captured by a Spanish privateer off the capes of Virginia.[64] In 1753 a fifty-five-ton whaling vessel, commanded by Captain Christopher Beetle and owned by John Norton of Martha's Vineyard, was "cast away on the coast of Carolina" during her second voyage. Norton chartered a vessel to rescue the ship, but upon their arrival "on Carolina, his vessel was gone with her sails, rigging, and appurtenances," leaving Norton with an additional loss of £500 sterling for the salvaging party.[65] In May, 1754, a severe storm along the Virginia coast did great damage to several vessels, including a Nantucket whaling ship, which lost its mast and two crew members.[66]

Despite such problems, the American whaling industry continued to flourish in the decades prior to the Revolution, and the North Carolina whale fishery remained active throughout that time. Governmental interest in whaling is seen in the instructions to Governor Arthur Dobbs, who in 1754 was reminded that he should encourage the whale fishery and should make no claims that he was entitled to seize the profits of such whaling.[67] In 1754 Dobbs wrote that the previously uncharted safe harbor at Cape Lookout was being used by "our Whale fishers" during the winter months.[68] In the spring of 1755, Dobbs reported that he had just returned from an inspection trip to Cape Lookout, "where the whale fishers from the Northward have a considerable fishery from Christmas to April, when the whales return to the northwd."[69]

[62]Richard C. Kugler, "The Whale Oil Trade, 1750-1775," *Publications of the Colonial Society of Massachusetts*, LII (1980), 153-173.

[63]Goode, "Whales and Porpoises," 8.

[64]Starbuck, "American Whale Fishery," 171.

[65]Starbuck, "American Whale Fishery," 42, 171.

[66]*Boston Evening Post*, May 20, 1754.

[67]Saunders, *Colonial Records*, V, 1135.

[68]Saunders, *Colonial Records*, V, 159.

[69]Saunders, *Colonial Records*, V, 346.

WHALE ON BEACH AT BEAUFORT

A local crew harpooned and subdued this whale shown beached on Shackleford Banks in the 1890s. Engraving from *Bulletin of the North Carolina Department of Agriculture*, 14 (April, 1894), 4.

Although Governor Dobbs's statements indicated an active whale fishery, not all New Englanders were successful in Carolina waters. After returning from a visit to Wilmington in 1752, Ashley Bowen of Massachusetts was hired by "a company of whalemen to engage my vessel, for me to go [as] master." Bowen departed from Cape May in his sloop *Susannah* in late November, 1753, with "12 men and all their appurtenances for whaling and two whaleboats and set out for a whaling voyage" to North Carolina. Passing Currituck on December 8 and Hatteras and Ocracoke on the ninth, Bowen's sloop arrived at Cape Lookout harbor on December 10, where they found a sloop from Nantucket under the command of Benjamin Bunker. Later in the afternoon two more whaling sloops arrived from Nantucket, commanded by John Starbuck and a Captain Macy, both bound for Cape Fear "to try another berth for whaling at the south." After several unsuccessful days in the Lookout area, Bowen headed south past Bear Inlet and New River, arriving at Brunswick and Wilmington on December 17. From there he proceeded to Lockwood's Folly, where "our people went on shore . . . to build them houses," suggesting that they would establish a base of operations there for the season. On December 23, Bowen accompanied a Captain Lyon to Little River, where they had breakfast with some of the whalemen and stayed on board a Rhode Island sloop anchored there. Although a whale had been seen at Lookout on December 13, Bowen and his acquaintances had little luck that season. "I tarried in Cape Fear River all winter. The

CUTTING BLUBBER.

Whalers profited by selling the oil of whales. In these drawings Shackleford whalers extract the oil by removing the blubber (above) and then boiling it in a process called "trying out" (below). From *Bulletin of the North Carolina Department of Agriculture*, 14 (April, 1894), 5, 7.

"TRYING OUT" OIL.

whalemen at Lockwood's Folly got nothing," while his crew on the *Susannah* "did not get a drop of oil."[70]

Other evidence of whaling just before the Revolution includes the 1757 sale of two tracts of land from John Shackleford to Joseph Morse and Edward Fuller, both of whom were apparently involved in whale fishing. The conveyed properties were on the beach between Topsail Inlet and

---

[70]Philip Chadwick Foster Smith (ed.), "The Journals of Ashley Bowen (1728-1813) of Marblehead," *Publications of the Colonial Society of Massachusetts*, XLIV (1973), 33-41.

Cape Lookout, and the sale included the "privileges to Point Lookout Bay . . . to have liberty to fish and whale in said Bay."[71] In 1762 a vessel transporting Moravians to North Carolina passed near Cape Lookout and sighted the remains of butchered whales. "We lay so close to land," wrote one of the brethren on May 9, "that we might have gone ashore, but it was only a stretch of sand, surrounded by water. We saw on it the bones of whales, which had been caught in the neighborhood, and cut up there." In 1764 the council at Wilmington took note that among the laws passed in the previous session of Parliament was an act for the "encouragement of the Whale fishery" in the colonies.[72] In September, 1764, Captain Jacob Lobb in His Majesty's sloop *Viper* surveyed Cape Lookout and noted on his map the location of "Whaler's Hutts" on the east end of what is now Shackleford Banks at a point almost due north of the tip of Cape Lookout.[73] In March of the following year, a French traveler visited Cape Lookout and observed that "there were some whale fishers tents" located there.[74] Some years later Henri Louis Duhamel du Monceau reported that British colonial whaling activities had been centered in three areas, New England, New York, and "Carolina, where they maintain many vessels, which carry the product of their fishery to England."[75]

Port records in the decade immediately preceding the Revolutionary War also reveal local trade actions involving whale products, although some may have involved transshipments from outside North Carolina. In July, 1764, a vessel with spermaceti candles cleared Port Brunswick for London, and in 1767 another ship bound from Brunswick to the Bay of Honduras included oil in its cargo.[76] Total exports from Port Beaufort in 1768 amounted to 1,126 gallons of whale oil and 150 pounds of whalebone.[77] From 1763 to the Revolution, ships belonging to Aaron Lopez of Newport, Rhode Island, conducted an active trade between the two colonies; and with Lopez's interest in whaling and candles, considerable movement of whale products occurred thereby.[78] Finally, on January 30, 1775, the *Lucy*, a twenty-five ton schooner with Captain Asa Hatch,

---

[71]Carteret County Deeds, Book F, p. 456, microfilm copy, State Archives; Charles Livingston Paul, "Colonial Beaufort: The History of a North Carolina Town" (unpublished master's thesis, East Carolina University, Greenville, 1965), 97, hereinafter cited as Paul, "Colonial Beaufort."

[72]Adelaide L. Fries, Douglas LeTell Rights, Minnie J. Smith, and Kenneth G. Hamilton (eds.), *Records of the Moravians in North Carolina* (Raleigh: North Carolina Historical Commission, 11 volumes, 1922-1969), I, 258; Saunders, *Colonial Records*, VI, 1081.

[73]"A Plan of the harbour of Cape Lookout surveyed and sounded by His Majesty's sloop Viper, Captain Lobb in Sepr. 1764. . . ," manuscript map, Howe Collection No. 19, Geography and Map Division, Library of Congress, Washington, D.C.

[74]"Journal of a French Traveller in the Colonies, 1765," *American Historical Review*, XXVI (July, 1921), 733.

[75]Henri Louis Duhamel du Monceau, *Traité général des pesches* (Paris: Saillant and Nyon, 4 volumes, 1769-1782), IV, 10; Frederick W. True, "The Whalebone Whales of the Western North Atlantic," *Smithsonian Contributions to Knowledge*, XXXIII (1904), 44, hereinafter cited as True, "Whalebone Whales."

[76]Treasurers and Comptrollers Papers, Ports, State Archives.

[77]Kell and Williams, *Carteret County*, 107.

[78]Virginia Bever Platt, "Tar, Staves, and New England Rum: The Trade of Aaron Lopez of Newport, Rhode Island, with Colonial North Carolina," *North Carolina Historical Review*, XLVIII (January, 1971), 1-18.

sailed from Port Roanoke bound for Salem, Massachusetts, with whale blubber as cargo.[79]

This thriving American industry was devastated by the Revolutionary War. In February, 1775, a bill was introduced into Parliament prohibiting the colonies from carrying on any fishery on the banks of Newfoundland or any other part of the North American coast. The ensuing years of embargoes, seizures, war, destruction of ships, and port blockades decimated the American whaling enterprise. Just before the Revolution, the industry employed some 4,700 men and 360 vessels, with an annual production of 45,000 barrels of sperm oil, 8,500 barrels of whale oil, and 75,000 pounds of whalebone. By 1789 the fleet had been reduced to 130 vessels and the annual production of sperm oil to 10,000 barrels.[80]

The effects of the Revolution on North Carolina whaling are not well documented, although in at least one instance military activities came into conflict with local whale hunting. In 1776 David Wade petitioned the council and governor for release from what he considered to be an unjust extension of his enlistment with the Core Sound Company, the independent militia in the area. Wade claimed that he had served out his six-month tour, but Captain Enoch Ward, commander of the company, had arrested Wade for desertion when Wade left to join one Captain Pinkum "to go a whaling" along "the banks."[81] Shore whaling was still active in April, 1782, when a group of whalers were noted near the ruins of Fort Hancock on Shackleford Banks, where they waited a chance to pursue their quarry.[82]

After the Revolution, port records also contain evidence that whale products were common items on ships clearing North Carolina. Although spermaceti candles and some of the oil barrels could have been transshipments from outside the state, most of the outbound oil was probably from whales processed locally. Not surprisingly, the principal scene of that export business was Port Beaufort, where between April, 1785, and December, 1789, at least eleven sloops and schooners cleared with "oil" as cargo bound for such locales as Rhode Island, Baltimore, Philadelphia, New York, Boston, England, the West Indies, and Guadeloupe. From October, 1784, to June, 1789, Port Currituck was the point of origin for eight outbound ships carrying oil, mostly to Baltimore and Virginia. Port Brunswick cleared five ships between January, 1787, and December, 1789, carrying oil to Dublin, Charleston, and the West Indian islands of St. Barthelemy, St. Eustatius, and Martinique; the brig *Polly* included spermaceti candles at her departure from Brunswick in April, 1788. Less intense activity was recorded at Port Bath and Port Roanoke, where single shipments of oil were noted in 1785 and 1786 respectively.[83]

---

[79]Treasurers and Comptrollers Papers, Ports, State Archives.

[80]Clark, "Whale Fishery," 65-67, 116-131; Starbuck, "American Whale Fishery," 59-77.

[81]Walter Clark (ed.), *The State Records of North Carolina* (Winston and Goldsboro: State of North Carolina, 16 volumes, numbered XI-XXVI, 1895-1906), XXII, 894-895.

[82]Kell and Williams, *Carteret County*, 21.

[83]For the names of vessels transporting whale products from North Carolina see table 1.

TABLE 1
Vessels Transporting Whale Products from North Carolina, 1785-1789

Vessel	Weight	Captain	Destination	Port of Departure	Date of Departure	Cargo
Sloop Nancy	20 tons	John Groton	Rhode Island	Beaufort	April 27, 1785	Whalebone
Schooner Nancy	12 tons	Solomon Fuller	Baltimore	Beaufort	May 22, 1787	Oil
Schooner Raven	18 tons	Warback	Swansborough	Beaufort	February, 1788	Oil
Sloop Industry	30 tons	Hubble	Philadelphia	Beaufort	May 31, 1788	Oil
Sloop New York Packet	60 tons	Griffin	New York	Beaufort	June 16, 1788	Oil
Sloop Friendship	65 tons	Johnston	Bath	Beaufort	February 17, 1789	Oil
Sloop Charlotte	18 tons	Samuel Chadwick	Boston	Beaufort	April 17, 1789	Oil
Schooner Polly	65 tons	Turner	West Indies	Beaufort	April 25, 1789	Oil
Schooner Betsey	80 tons	Smith	England	Beaufort	May 2, 1789	Oil
Schooner Fanny	60 tons	Benjamin Leecraft	Guadeloupe	Beaufort	May 7, 1789	Oil
Schooner Active	67 tons		Roanoke	Beaufort	December 29, 1789	Oil
Ship Minerva	100 tons	Gideon Freeborn	Dublin	Brunswick	January 13, 1787	Oil
Schooner Wilmington Packet		Luke Swain	Charleston	Brunswick	June 30, 1787	Oil
Brig Polly	30 tons	Edmund Case	St. Bartholomew	Brunswick	April 17, 1788	Spermaceti Candles
Schooner Good Hope		Henry Hunter	St. Eustatius	Brunswick	December 5, 1788	Oil
Schooner Sally	38 tons	Thomas Potter	Martinique	Brunswick	December 1, 1789	Oil
Lively		John Litchfield	Baltimore	Currituck	October 17, 1784 September 15, 1785	Oil
Sally	5 tons	Caleb Chaplan	Baltimore	Currituck	June 19, 1786	Oil
Nancy	15 tons	Solomon Ashby	Baltimore	Currituck	November 28, 1788	Oil
Sally	9 tons	William Arthur	Baltimore	Currituck	January, 1789	Oil
Industry	12 tons	John Cudworth	Baltimore	Currituck	April 17, 1789	Oil
Polly	5 tons	Mark Davis	Richmond	Currituck	May 20, 1789	Oil
	5 tons	William Price	Virginia	Currituck	June 30, 1789	Oil
Schooner Phoenix	60 tons	John Barry	St. Bartholomew	Roanoke	March 4, 1786	Oil

SOURCE: The above list was compiled by Dr. Wilson Angley, researcher, Division of Archives and History, from the Treasurers and Comptrollers Papers, Ports, State Archives. In 1785 five casks of oil were shipped out on an unnamed brig from Port Bath. Although some of that oil may have been from dolphins and fish, most was probably produced from whales. The spermaceti candles on the brig *Polly* were probably transshipments from New England.

In the interval between the Revolution and the War of 1812, American whaling slowly regained much of the dominance lost during the war for independence.[84] Local whaling continued in North Carolina, although documentation is sparse. From May to October, 1806, federal official William Tatham explored the coast from Cape Hatteras to Cape Fear, and his "Separate Report" reveals that he conducted most of that work from a whaleboat rented for the purpose. Tatham also noted that the mullet fishermen of "Cart Island" and "Middle Island" near Beaufort participated in whaling during the winter: "I understand in the winter they are partially employed in whaling on Cape Lookout."[85] Some years later, in December, 1810, Jacob Henry, Carteret County legislator, wrote to Thomas Henderson, Jr., publisher of the Raleigh *Star*, that "Something is done every year in the Whale fishery" at Beaufort.[86]

A somewhat different form of marine mammal fishing emerged during the late eighteenth and early nineteenth centuries with the development of an industry for producing oil from the bottlenose dolphin. Dolphins, or "porpoises" as they were commonly called, had been captured on the Carolina coast for oil as early as Lawson's time, but an organized industry in North Carolina may have begun with the efforts of North Carolina merchant and politician John Gray Blount and his business associate John Wallace.[87] In the summer of 1790, Wallace wrote to Blount regarding porpoises and "porpus Seins," which were being used to capture the creatures at Hatteras. In 1793 Blount and Wallace were using their main lighter vessel, the *Beaver*, in the porpoise industry, which by 1802 was providing the oil for the lighthouse lamp at Shell Castle Island. In 1800 Wallace was able to inform Blount that "Our porpouse fishing I am told is doing very well," and in January, 1803, Wallace notified Blount that he had procured thirty oil barrels and sent them to Hatteras, where they would soon be needed for the beginning of the season's fishery.[88] The end of the Blount-Wallace ventures at Shell Castle Island did not lessen the

---

[84]Starbuck, "American Whale Fishery," 77-96; Clark, "Whale Fishery," 140-141.

[85]"The Separate Report of William Tatham one of the Commissioners appointed to survey the Coast of North Carolina from Cape Hatteras to Cape Fear . . . to Honl. Albert Gallatin, Secy. of the Treasury &c., Jany., 1807," manuscript report, Records of the Coast and Geodetic Survey, Record Group 23, National Archives, Washington, D.C., 8, 11, 17, 22-24, 26, 44, 47, 51-52, hereinafter cited as Tatham, "Separate Report of William Tatham," Records of Coast and Geodetic Survey, RG 23; G. Melvin Herndon, "The 1806 Survey of the North Carolina Coast, Cape Hatteras to Cape Fear," *North Carolina Historical Review*, XLIX (July, 1972), 242-253.

[86]A. R. Newsome, "A Miscellany from the Thomas Henderson Letter Book, 1810-1811," *North Carolina Historical Review*, VI (October, 1929), 399, hereinafter cited as Newsome, "Thomas Henderson Letter Book."

[87]Lawson, *New Voyage to Carolina*, 158; John Gray Blount and John Wallace organized an ambitious and diversified commercial venture at Shell Castle Island between Portsmouth and Ocracoke. Alice Barnwell Keith, "Three North Carolina Blount Brothers in Business and Politics, 1783-1812" (unpublished doctoral dissertation, University of North Carolina, Chapel Hill, 1940), 1-470.

[88]John Wallace to John Gray Blount, August 26, September 6, 1790, Alice Barnwell Keith, William H. Masterson, and David T. Morgan (eds.), *The John Gray Blount Papers* (Raleigh: Division of Archives and History, Department of Cultural Resources, 4 volumes, 1952-1982), II, 98, 106-107, hereinafter cited as Keith and others, *Blount Papers*; John Wallace to John

The *Amelia* (above) of New Bedford was similar to the northeastern schooners that engaged in whaling off North Carolina in the late nineteenth and early twentieth centuries. The vessel's deck and interior plans appear on the adjacent page. "Trying out" pots stand just left of midship in both diagrams. Barrels for storing whale oil are pictured in the interior view. Drawings from George Brown Goode (ed.), *The Fisheries and Fishery Industries of the United States* (Washington: Commission of Fish and Fisheries, 5 sections, 1884-1887), V, plates 186, 188.

interest in dolphin catching on the coast, for the industry was continued by others up until the outbreak of the Civil War.[89]

In 1806 Tatham reported that "The inhabitants [of Beaufort] also carry on the Porpoise . . . fisheries jointly wih the People of . . . Core Sound, at Cape Lookout, where about two hundred barrels of oil were produced last year."[90] In 1810 Jacob Henry informed Thomas Henderson, Jr., that the porpoise industry at Beaufort involved "much more" activity than even the local whaling, with the porpoise oil bringing about 40 cents per

---

Gray Blount, March 24, 1800, Keith and others, *Blount Papers*, III, 351; John Wallace to John Gray Blount, January 10, 1803, Keith and others, *Blount Papers*, IV, 5; Sarah Olson, *Historic Resource Study: Portsmouth Village* (Denver: National Park Service, 1982), 20, 55, 61.

[89]James G. Mead, "Preliminary Report on the Former Net Fisheries for *Tursiops truncatus* in the Western North Atlantic," *Journal of the Fisheries Research Board of Canada*, 32 (1975), 1155-1162, hereinafter cited as Mead, "Net Fisheries for *Tursiops truncatus*"; Earll, "North Carolina Fisheries," 490.

[90]Tatham, "Separate Report of William Tatham," Records of Coast and Geodetic Survey, RG 23, p. 25.

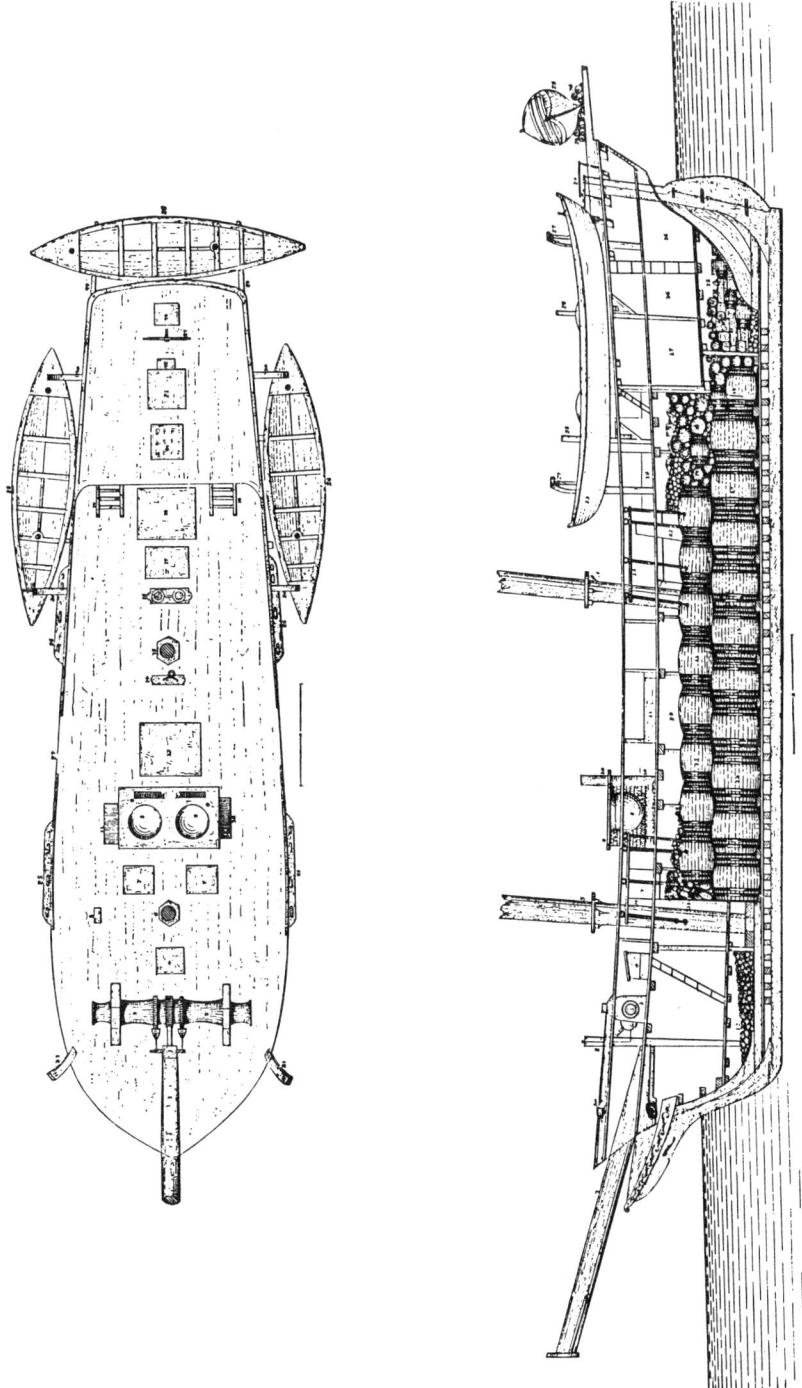

gallon.[91] By 1810 the dolphin or porpoise fishery was being conducted between Cape Hatteras and Bear Inlet, a stretch of coast where "often immense herds of them" could be observed moving close by the shore.[92]

From 1810 to 1860 the dolphin fishery at Beaufort was conducted regularly each year by one to three crews. Fishermen took the dolphins with heavy seines about 800 yards long, which, because of their weight and bulk, were handled in sections of 200 yards each. Crews shot the seines, with the ends securely fastened, simultaneously from four boats. The two outer seines were lashed together; and at the signal from a lookout, the seines were fired toward shore in the form of a semicircle to entrap the herd of dolphins. The men then hauled in the seine as close as convenient toward shore and then landed the dolphins with smaller and stouter nets. The larger seine consisted of 11-inch mesh, while the smaller shore nets had about 9-inch mesh. Although the dolphins seldom tried to break through the nets, some occasionally jumped over the cork line, an event that often resulted in the remaining dolphins following to escape. Each crew usually consisted of fifteen to eighteen men, and the season lasted from late December until the first and rarely the fifteenth of April. An average season's catch was 400 to 500 dolphins, and about five to six dolphins were needed to produce each barrel of oil.[93]

Information on local whaling during the first half of the nineteenth century consists mainly of oral traditions that were recorded by scientists and historians in the years after the Civil War. The Beaufort shore whaling industry was presumably continued through most of that period, as the oldest residents of the area claimed in 1880 that local whaling had begun prior to their earliest recollections, and later writers stated that at least two generations had engaged in whaling prior to the time of James Lewis, a local whaler who was born in 1830. At Bear Island, just west of Bogue Bank, a Captain Daniel Heady reportedly established a whaling station in the early 1800s and later became involved with porpoise fishing there. This would indicate that shore whaling had been pursued at least fairly regularly from the time Tatham and Henry noted the activity in 1806 and 1810.[94]

Whaling vessels from New York and New England also continued their activities along the Carolina coast in those decades between the War of 1812 and the Civil War. During that interval, Yankee whalers, having largely depleted the commercially desirable cetaceous populations of the Atlantic, extended their voyages into the Pacific, Indian, and western Arctic oceans, and in so doing achieved international domination of their industry during the 1840s, in what was to be known as the "Golden Age" of American whaling.[95] Details of pelagic whaling in North Carolina

[91]Newsome, "Thomas Henderson Letter Book," 399.

[92]Earll, "North Carolina Fisheries," 490.

[93]Earll, "North Carolina Fisheries," 490.

[94]Holland, *Survey History of Cape Lookout*, 13; Marcia Constantino, "Hammocks Beach State Park," *Wildlife in North Carolina*, 41 (June, 1977), 27; Stick, *Outer Banks*, 185-186; Earll, "North Carolina Fisheries," 490; Odum, *H. H. Brimley Writings*, 111.

[95]Sanderson, *Follow the Whale*, 248.

waters are sparse during much of that period, but more than forty vessels from northeastern ports sailed in the Hatteras ground or along the Outer Banks during the 1830s and 1840s. These activities ranged from close by the shoreline or major capes to far at sea, within and along the Gulf Stream and the edge of the Continental Shelf and beyond. In some instances, the whalers made only a quick transit through the area, perhaps bound to or from their home ports in New England, but other vessels spent much of their entire voyage whaling the Carolina waters.[96]

Later writers, such as Captain N. E. Atwood of Provincetown, recalled that "In 1837 the *Edward and Rienzi* . . . went . . . toward Cape Hatteras. No whaling vessels had ever been there before, and she found sperm whales abundant, and since that time the Hatteras ground and the Charleston ground . . . have been favorite cruising grounds for the Provincetown fleet."[97] Extant records provide little supporting documentation of the particulars of Atwood's claim, although the Provincetown whalers apparently outnumbered all others in Carolina waters in the 1840s and 1850s.[98]

During the 1830s, however, Massachusetts vessels of Rochester and New Bedford registry were frequent visitors to offshore waters of the North Carolina coast. New Bedford whalers sailing through the area included the *Delight, Franklin, Hydaspe, Juno, Laurel*, and the *Rising States*. The *Charleston Packet*, a 184-ton New Bedford brig commanded by Ebenezer Ellis, Jr., captured pilot whales in May, 1837, at 36° north latitude and 73° west longitude, had "a whale along side" on May 8, 1838, at 35°39' north latitude and 74°25' west longitude, and returned to the area again in May of 1839 and 1840. A veteran of three decades of whaling, the *Winslow*, a 263-ton ship of New Bedford, with Captain Gifford in charge, passed "off Cape Fear April 10," 1838, and by May 20 had taken "a 90 bbl whale along side." Whalers from Rochester included the *Sarah, LeBaron, Mattapoisett, Willis, Lagrange*, and *Solon*, while vessels noted from other ports were the *Ann Maria, William*, and *Taunton* of Fall River, the *Brunette* of Falmouth, the *Mexico* and *Thomas Winslow*

---

[96]Dennis Wood, "Abstracts of Whaling Voyages [1835-1875]," an unpublished manuscript, 5 volumes, n.d., New Bedford Free Public Library, New Bedford, Massachusetts, hereinafter cited as Wood, "Abstracts of Whaling Voyages."

[97]N. E. Atwood, "Whale Fishery of Provincetown," in George Brown Goode (ed.), *The Fisheries and Fishery Industries of the United States* (Washington: Commission of Fish and Fisheries, 5 sections, 1884-1887), Section V, Volume II, 144-145, hereinafter cited as Atwood, "Whale Fishery of Provincetown." Atwood may have been referring to the schooners *Edwin* and the *Rienzi*, as there are no records of voyages or extant logbooks by Starbuck, Wood, or Downey and Adams of ships bearing the name *Edward and Rienzi*. Starbuck lists sixteen whaling voyages in the North Atlantic and Gulf of Mexico by the 115-ton *Rienzi*, spanning the period 1844 to 1863, when the vessel was captured and burned by a Confederate privateer. Starbuck, "American Whale Fishery," 414-585. The *Edwin* was a 100-ton Provincetown schooner that conducted four voyages, one each year from 1844 until 1848, when it was "withdrawn" from the industry. Starbuck, "American Whale Fishery," 414-415, 428-429, 438-439, 448-449. No logbooks are listed for these two vessels in Judith M. Downey and Virginia M. Adams (eds.), *Whaling Logbooks and Journals, 1613-1927: An Inventory of Manuscript Records in Public Collections* (New York: Garland Publishing Co., 1986), hereinafter cited as Downey and Adams, *Whaling Logbooks and Journals*; Clark, "Whale Fishery," 8-9, 24.

[98]Wood, "Abstracts of Whaling Voyages."

of Westport, the *Bruce* of East Haddam, the *Crawford* of Warren, the *Governor Hopkins* and the *Troy* of Bristol, the *Primrose* of Nantucket, and the *William and Joseph* of Holmes's Hole.[99]

By the 1840s, the Provincetown fleet began to dominate pelagic whaling off the Carolina coast, where pilot[100] and sperm whales were sometimes taken by vessels such as the schooners *Rienzi, Council, Belle Isle, Walter Irving, John Adams, William Henry, Tarquin, Louisa,* and *Stranger*; the bark *Fairy*; and the brigs *Lewis Bruce, Gem, Cadmus, Janes Howes, Carter Braxton, Samuel Cook,* and *Franklin.* The *Edwin,* a 100-ton schooner owned by Samuel Cook, was "supposed out on [a] blackfishing cruise" to the Carolina coast in the spring of 1844, but on June 13 she was "20 miles N of Hatteras with a whale alongside" and her cargo upon returning in July consisted of 300 barrels of sperm whale oil and only twenty barrels of blackfish oil. Subsequent visits by the *Edwin* around Cape Hatteras yielded whales in the spring of 1845, 1846, and 1847. The *Grand Island,* another 100-ton schooner also owned by Samuel Cook, fared less well, being forced to abandon her cruising in the area in May, 1846, and to put into Hampton Roads because of an outbreak of smallpox, the mate and two crew members being afflicted. Another of Samuel Cook's schooners, the *Chanticleer,* cruised the coast in 1849 and was off "Currituck" on June 1, 1850; and the schooner *Medford* passed "off Hatteras bound home" on May 25, 1846.[101]

Provincetown vessels were not the only whaling ships off the North Carolina coast during the 1840s. Whalers from other ports venturing through the area included the *Leonidas* of Fall River, the *Edward* and the *Sarah* of Mattapoisett, the *Peru* of New Bedford, and the *Juno,* the *President,* and the *Barclay* of Westport. The *Pavilion,* a 150-ton brig of Edgartown, "passed off Cape Look Out" on March 29, 1842, and returned to the Hatteras area on April 5 after a brief sojourn by Cape Charles. Other whalers plying the area included the *Exchange* of Plymouth, the *Two Sisters* of Sippican, the *Wickford* of Sag Harbor, and the *Henry* of Stonington. Some fared badly off the North Carolina coast, as did the *Young Eagle,* a 379-ton whaling ship built in 1832 at Rochester and owned by Simeon Starbuck of Nantucket. Returning from a four-year cruise in the Pacific, it was lost on November 2, 1847, at 35°5′ north latitude and 73°35′ west longitude.[102]

By the 1850s the continued decline of whales in the Atlantic and the pursuit of bowhead whales in the western Arctic forced more of the American fleet into the Pacific basin, leaving the North Carolina coast

---

[99]Wood, "Abstracts of Whaling Voyages," I, 45, 56, 68, 91, 116, 134, 164, 210, 224, 268, 293, 300, 335-336, 416, 438, 463, 485, 490, 508, 515, 521, 531, 533, 544, 554.

[100]Pilot whales or blackfish (*Globicephala macrorhyncha*) are often involved in mass strandings along the coast of the western North Atlantic. Caldwell and Golley, "Marine Mammals," 27-28.

[101]Wood, "Abstracts of Whaling Voyages," I, 100, 368; II, 66, 116, 123, 203, 214, 232-233, 265, 314, 399, 410, 465, 481, 579, 633, 660.

[102]Wood, "Abstracts of Whaling Voyages," I, 268, 497, 530, 552, 558, 560, 569, 571, 577-578; II, 204, 538, 604, 683.

mostly to the Provincetown whalers who had dominated the area in the 1840s. Among the latter were small vessels, many suited for cruises of less than one year's duration, such as the *E. Nickerson, Franklin, Montezuma, Rienzi, Richard, R. E. Cook, V. Doane*, and the *Chanticleer*. New Bedford vessels largely operated elsewhere. The barks *Orray, Taft*, and *Sea Flower* were among the only New Bedford craft to venture into North Carolina waters, whose importance to the Yankee whaling industry had clearly diminished. Only a few other whalers used the area in the 1850s, such as the *Monteray* of Edgartown, the *Afton* of Boston, the *Palmyra* of Nantucket, and the *Sarah* and the *Samuel and Thomas* of Mattapoisett.[103]

By the time of the Civil War, the "Golden Age" of American whaling had passed. Factors contributing to the slow but inexorable decline of the industry included the discovery of petroleum, the rising cost of outfitting ships, and the need for lengthy and dangerous voyages to remote seas, because the Atlantic waters had been largely depleted of easy prey.[104] The war itself further exacerbated the situation, for Confederate war ships, such as the *Alabama* and the *Shenandoah*, extracted a heavy toll on Yankee whalers, who were hounded even into the northern Pacific and western Arctic oceans.[105] Through most of the war, northern whaling vessels prudently avoided the Carolina coast, where only a few ships, such as the *Falcon* of Salem and the *Montezuma* of Provincetown, chose to venture.[106]

With the cessation of hostilities, the dual pattern of local shore whalers and northern ocean whalers resumed in coastal North Carolina, although the New England fishermen were greeted with considerable hostility by the natives. Most Yankee whalers remained well offshore, and the volume of activity was far less than during the three decades immediately preceding the war. A few New Bedford ships, such as the *John W. Dodge, Tropic Bird*, and *Union*, cruised the area, but many of the pelagic whalers passing off North Carolina originated from ports that had minimal involvement with the Carolina coast prior to the war. From Fairhaven sailed the *William and Henry, Washington Freeman, Union*, and *Ellen Rodman*; the last, for example, noted its position "off Hatteras" in August, 1870. A Boston fleet of whalers owned by Herman Smith was active along the southeast coast and included the *William Martin*, the *Sarah Lewis*, and the *Herman Smith*. Vessels of Sippican registry were the *Graduate* and the *Herald*. The latter successfully captured sperm whales off Hatteras from June through September, 1867. From Newburyport the *Georgia* was

---

[103]Wood, "Abstracts of Whaling Voyages," I, 42; II, 116, 233, 242, 610; III, 238, 272, 399, 459, 477-478, 481-482, 485-486.

[104]Starbuck, "American Whale Fishery," 97-100, 109-114; Sanderson, *Follow the Whale*, 250-253; Clark, "Whale Fishery," 70-73.

[105]The *Shenandoah*, a "rebel" privateer steamer, pursued its Yankee victims into the North Pacific and Arctic, eventually destroying over thirty New England whaling ships. The *Alabama* employed the particularly sinister tactic of entrapping whalers as they responded to a burning ship. Starbuck, "American Whale Fishery," 100-103; Sanderson, *Follow the Whale*, 253.

[106]Wood, "Abstracts of Whaling Voyages," III, 485; IV, 301.

"on Hatteras ground" in June, 1869, and the *Hannah Grant* was cruising there from April to June, 1867. Other New England whalers along the Carolina coast during the late 1860s included the *Cohannet* of Marion and the *Thriver* of Beverly.[107] Although such pelagic activities were clearly on the wane after the Civil War, it was during this final period of North Carolina whaling that the most detailed accounts have survived concerning the local shore-based industry and its methods.

Soon after the Civil War, physician and naturalist Elliott Coues was assigned duty as an army doctor at Fort Macon in Carteret County. While there between February, 1869, and November, 1870, he recorded his observations on the natural history of the region in his letters and a number of publications.[108] Coues noted that the natives were mostly fishermen, to whose "regular employment with the seine they add the capture of an occasional whale in spring."[109] Coues and his successor at Fort Macon, Dr. Henry C. Yarrow,[110] also recorded the pilot whale or blackfish: "a single specimen observed . . . taken at a porpoise-fishing on Shackleford banks, six miles from Fort Macon." Coues and Yarrow were informed by the fishermen that the blackfish was "rarely seen or captured."[111]

Coues also claimed that two species of dolphins could be observed in the area, both of which were taken at Shackleford Banks, where they were captured for their oil using "nets made especially for the purpose." He noted that the less common of the two species was "smaller and darker than the other." The dolphins were present all year but "most abundant in spring and fall, during the migration of the fish upon which they prey."[112] Unfortunately, neither Coues nor Yarrow defined the identity of these two species, but in 1871 Yarrow sent to the Smithsonian Institution a number of dolphin skulls that were later identified as belonging to the bottlenose dolphin.[113]

---

[107]Wood, "Abstracts of Whaling Voyages," IV, 110, 206, 208, 233-234, 242, 246, 277, 303, 348, 361, 363-364, 365.

[108]Elliott Coues (1842-1899), zoologist, historian, mystic, and army surgeon, spent much of his time at Fort Macon engaging in fieldwork and writing technical and popular papers and books. Paul Russel Cutright and Michael J. Brodhead, *Elliott Coues, Naturalist and Frontier Historian* (Urbana: University of Illinois Press, 1981), 105-120.

[109]Elliott Coues, "Fort Macon, North Carolina," in John S. Billings (ed.), *Report on Barracks and Hospitals* (Washington: War Department, Surgeon General's Office [Circular No. 4, December 5] 1870), 90.

[110]Henry Crecy Yarrow (1840-1929), army surgeon, naturalist, and native of Philadelphia (where he received his doctor of medicine degree from the University of Pennsylvania), remains known for distributional studies of birds, mammals, reptiles and amphibians. Edgar Erskine Hume, *Ornithologists of the United States Army Medical Corps* (Baltimore: Johns Hopkins University Press, 1942), 533-549.

[111]Elliot Coues and H. C. Yarrow, "Notes on the Natural History of Fort Macon, N.C., and Vicinity (No. 4)," *Proceedings of the Academy of Natural Sciences of Philadelphia*, XXX (1878), 22.

[112]Elliott Coues, "Notes on the Natural History of Fort Macon, N.C., and Vicinity (No. 1)," *Proceedings of the Academy of Natural Sciences of Philadelphia*, XXIII (May, 1871), 13-14, hereinafter cited as Coues, "Natural History of Fort Macon."

[113]Mead, "Net Fisheries for *Tursiops truncatus*," 1157.

Physician and naturalist Elliott Coues recorded his observations of whales and whaling while serving as an army doctor at Fort Macon in Carteret County soon after the Civil War. Photograph from *Dictionary of American Portraits* (New York: Dover Publications, 1968), 135.

Coues further noted the local whaling activities in the area in his correspondence with Spencer F. Baird,[114] assistant secretary at the Smithsonian:

Remembering your desire regarding cetacean remains, I have just made a boat trip several miles to the other side of Shackleford island to look at a whale caught yesterday . . . Right whale about 45 feet long—all the bones are at disposal of self or anybody else but it would be an augean job to clean them—besides these fresh bones, there are several well bleached jaws, humeri, vertebrae &c &c lying all about. . . . Fishermen talk of only three whales about here—"Right," "Scrag," & "Humpback."[115]

[114]Spencer Fullerton Baird (1823-1888), zoologist and assistant secretary of the Smithsonian Institution from 1850 to 1878, remains perhaps best known for promoting the scientific exploration of North America. W. H. Dall, *Spencer Fullerton Baird: A Biography* (Philadelphia: J. B. Lippincott, 1915).

[115]Elliott Coues to Spencer F. Baird, April 15, 1869, Incoming Correspondence, Assistant Secretary, Smithsonian Institution Archives, Washington, D.C., quoted by permission of Peter Coues.

Coues summarized his experiences with local whaling and whales in subsequent publications on the natural history of Fort Macon:

Balaena cisarctica . . . An individual, which I rather suppose than know to be of this species, was taken in May, 1869, off Shackleford. It measured about forty-five feet in length. The fishermen called it a "right whale." Besides this kind, they spoke of two others, that they occasionally captured, under the names of "Scrag" and "Humpback." . . . They usually take two or three each spring. Remains of whales . . . are strewn abundantly along the beach.[116]

Except for the right whale, which continued to be the principal target of the shore whalers, the identities of the species mentioned by Coues are not certain. Humpback whales might have been taken locally; later authors made similar claims, and there are modern incidents of stranded humpbacks on the sand banks.[117] However, apparently no actual records exist of the shore-based capture of humpbacks on the North Carolina coast, and thus Coues's informants may have been mistaken. Furthermore, the identity of the "scrag" whale is unclear; perhaps it was the Atlantic form of the gray whale or perhaps a so-called "dry skin" right whale, an emaciated specimen whose thin blubber and poor yield of oil usually made the animal undesirable for whalers.[118]

During the 1870s and 1880s the Beaufort whaling ventures continued to attract the attention of scientists. State geologist W. C. Kerr reported that whales were taken mostly along Shackleford Banks between Cape Lookout and Fort Macon, with the season principally in April and May. In some years as many as five or six right whales would be killed in a period of one or two weeks at the peak of the migration.[119] On one occasion, whalers killed two sperm whales near Cape Lookout, one of which measured 62 feet in length. Kerr claimed in 1875 that the value in oil and whalebone for a right whale usually ranged from $1,200 to $1,500,[121] and

[116]Coues, "Natural History of Fort Macon," 13.

[117]The humpback whale (Megaptera novaeangliae), nearly worldwide in distribution, occurs largely offshore in the North Atlantic, where it has been extensively hunted in the past. Edward Mitchell and Randall R. Reeves, "Catch History, Abundance, and Present Status of Northwest Atlantic humpback whales," Report of the International Whaling Commission, Special Issue No. 5 (1983), 153-212; Caldwell and Golley, "Marine Mammals," 29; Pilson and Goldstein, "Marine Mammals," 20-22.

[118]Although the gray whale (Eschrichtius robustus) occurred in the western North Atlantic ocean during early colonial days, the species was apparently exterminated from the region by the eighteenth century, leaving only the Pacific population surviving into modern times. Skeletal remains of gray whales have been found occasionally along the Outer Banks. James G. Mead and Edward D. Mitchell, "Atlantic Gray Whales," in M. L. Jones, S. L. Swartz, and S. Leatherwood (eds.), The Gray Whale (Orlando: Academic Press, 1984), 33-53.

[119]W. C. Kerr, Report of the Geological Survey of North Carolina [1866-1887] (Raleigh: State of North Carolina, 3 volumes, 1867-1888), I, 15, hereinafter cited as Kerr, Report of the Geological Survey.

[120]News and Observer, September 21, 1941; Pat Dula Davis, Kathleen Hill Hamilton, and Charles O. Pitts, Jr. (eds.), The Heritage of Carteret County North Carolina (Winston-Salem: Hunter Publishing Co., 2 volumes, 1984), II, 419, hereinafter cited as Davis, Hamilton, and Pitts, Carteret County.

[121]Kerr, Report of the Geological Survey, I, 15.

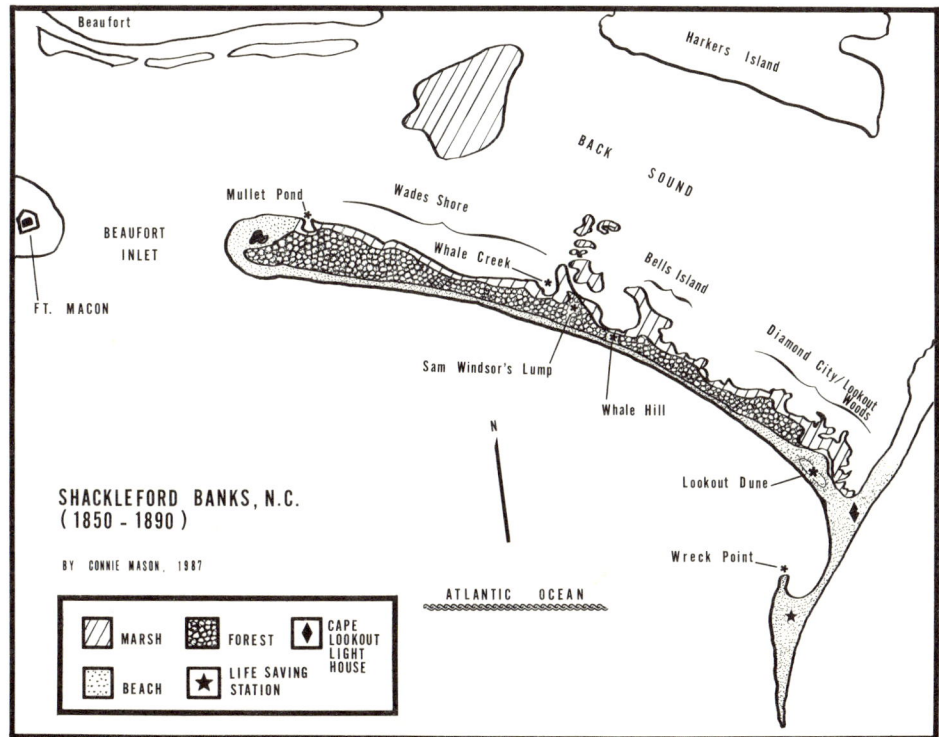

This map shows where a number of residents of Shackleford Banks lived and conducted fishing and whaling activities from 1850 to 1890. Map by Connie Mason, National Park Service.

in the 1880s another authority on whales, A. H. Clark, gave similar figures, stating that the typical annual catch of four whales would yield about $4,500 in earnings.[122] Although in the 1870s and 1880s, right whales were the principal targets, sperm whales were occasionally taken, but no specific account of capture has been found. The fin, humpback, and "scrag" whales were sometimes seen but seldom if ever chased or taken.[123]

Shore-based whaling activities apparently extended as far north as Cape Hatteras and southward to Little River,[124] although most whale catching was centered at the Beaufort area. The whalers organized themselves into "camps" among the sand hills along the shore of the barrier islands. Each crew posted lookouts and built a "house out of rushes in some desirable location near the shore, for protection against the

[122]Clark, "Whale Fishery," 40, 49.

[123]H. H. Brimley Papers, North Carolina Museum of Natural Sciences, Raleigh, hereinafter cited as Brimley Papers.

[124]Clark, "Whale Fishery," 49; Earll, "North Carolina Fisheries," 490.

weather." There they waited with their boats in readiness for the hunt. The number of men engaged in the whale fishery varied from year to year, usually consisting of two to three camps of about eighteen men each. In 1879 there were four camps with a total of seventy-two men, who captured five whales and shared in earnings of $4,000. In the spring of 1880 six crews totaling 108 men stationed themselves between Cape Hatteras and Bear Inlet. Most of the whales had already passed the coast before the crews organized, however, and only a single small whale was taken, yielding slightly more than $400 to the fishermen.[125]

A federal official, R. E. Earll, visited the region in April, 1880, and provided a detailed account of the methods employed by the Beaufort whalers:

The shore whalers resort to the outer beach with their boats and other apparatus about the 1st of February, and after building a camp for cooking and sleeping, they establish a "crow's-nest" or lookout station on one of the highest hills, where some of their number are stationed to watch for the whales that follow the shore in their migration toward the north. The season lasts until the 1st of May. A camp usually consists of three boat crews, of six men each, and while waiting for whales some of the men fish with seines for such fish as happen to be moving along the shore. A lookout is kept constantly in the crow's-nest, and when a whale comes in sight the signal is given and the boats start in pursuit. When the whale is overtaken the harpoon is plunged into it. A wooden drag is usually attached to the iron by means of a short line. This is at once thrown out, and the animal is allowed to "have its run." Harassed by the drag, the whale soon turns to fight, when the boats quickly overtake it, and one of the gunners shoots it with an explosive cartridge. When the creature has been killed it is towed to the shore, where it is cut up and the blubber tried out.[126]

Local whalers may have been successful during the 1870s, but two New England whaling vessels from Provincetown fared rather poorly at Cape Lookout. In the winter of 1874-1875, the *Daniel Webster*, a twenty-four-ton schooner, cruised off the cape for three months but eventually gave up and returned homeward without taking a single whale. Although A. H. Clark stated that the *Daniel Webster* was responsible for introducing the whaling gun at Beaufort,[127] local traditions claim that such muzzle-loading shoulder guns were made locally in the area or were introduced there by North Carolina whalers who purchased the guns in Baltimore.[128] The crew of the *Daniel Webster* may have been disappointed in failing to capture any whales, but another Provincetown whaler visiting the cape suffered a greater disaster.

---

[125]Earll, "North Carolina Fisheries," 490.

[126]Earll, "North Carolina Fisheries," 490.

[127]Clark, "Whale Fishery," 49; Earll, "North Carolina Fisheries," 490.

[128]The identity of the *Daniel Webster* is problematic. Starbuck lists two whaling ships bearing the name *Daniel Webster*, both over 300 tons weight and with no reference of their sailing in North Carolina waters or the North Atlantic during the years in question. Neither ship was of Provincetown registry. Starbuck, "American Whale Fishery"; Clark, "Whale Fishery"; Downey and Adams, *Whaling Logbooks and Journals*; Davis, Hamilton, and Pitts, *Carteret County*, I, 37; II, 417.

In the winter of 1878 the *Seychelle*, a forty-seven-ton whaling schooner captained by one E. Cook, arrived at the Beaufort area and commenced operations along the coast. By the summer of 1879, the crew had failed to kill a single whale, but the captain had been assured that the month of August was a fine time for sailing in those waters.[129] On the morning of August 18, however, the cape was lashed by a hurricane with winds estimated to exceed 165 miles per hour. The *Seychelle* was driven ashore near the United States Army Signal Corps Station at Cape Lookout, where it was perched about five feet above the high tide mark. Private H. J. Forman, on duty at the station, telegraphed a report of the incident to headquarters in Washington:

The whaling schooner Seychelle of Provincetown Mass 50 tons, Capt Cook, fishing in these waters was at anchor in the Hook parted her chains ran across wreck point, by almost superhuman efforts the crew got a small piece of canvas on her and the wind veering to the Southwest they succeeded in running her ashore within ½ mile of this station, where she now lies high and dry above the highest tide mark, a total loss. None of the crew were lost. At the time the vessel crossed wreck point she was drawing 12 feet of water, thus showing the tide to have been fuller than ever known at this place as the oldest inhabitants cannot remember of any tide ever before overflowing this point.[130]

Local residents recounted that the *Seychelle* was "stripped clean by wind and water," and the particular hurricane for many years thereafter was dubbed "old Cook's storm" in memory of the captain.[131] Although northern whalers from Provincetown and New Bedford continued to cruise the Hatteras ground and adjacent coastal waters, they apparently henceforth seldom visited the Cape Lookout area. Most of such activities were by fairly small vessels that chased sperm whales along the South American coast in the winter, moved to the Caribbean in January and February, entered the Gulf of Mexico in April, and then sailed northward in May by Cape Hatteras to the Hatteras ground, where they remained until September.[132]

Other New England whalers visited the North Carolina coast in pursuit of right whales, which were sought there as late as the 1880s.[133] For example, the whaling schooners *E. H. Hatfield* and *Emma Jane*, of Edgartown, Massachusetts, cruised the southeast coast in the late winter and early spring of 1881. The *Hatfield* anchored at Morehead City in early February, searched for whales along the coast between Fort Macon and Fort Caswell through April, and took on a crew at Wilmington on

---

[129]Stick, *Outer Banks*, 186; Earll, "North Carolina Fisheries," 490.

[130]Report of August 23, 1879, United States Army Signal Corps Reports from Cape Lookout, North Carolina (microfilm), Southern Historical Collection, University of North Carolina Library at Chapel Hill.

[131]Stick, *Outer Banks*, 186.

[132]Charles Haskins Townsend, "The Distribution of Certain Whales as Shown by Logbook Records of American Whaleships," *Zoologica*, XIX (April, 1935), 12; Clark, "Whale Fishery," 6, 8-9, 15, 22, 24, 144-145.

[133]Clark, "Whale Fishery," 15-16.

On this map are shown the major whaling grounds for the east coast of North and Central America. The area northeast of North Carolina's Cape Hatteras was known as the Hatteras ground. In that region pelagic whalers from the northeastern United States pursued sperm whales. From *Zoologica*, XIX (April, 1935).

May 1 before heading on to Bermuda a few weeks later. Only a single whale—a fin—was seen during the period, and the two vessels apparently never returned to whale along the North Carolina coast.[134]

Meanwhile, the dolphin industry had been revived by the early 1880s as an additional source of cetaceous oil and commerce. Dolphins had been so abundant at Hatteras Inlet in the winter of 1874-1875 that the waters "seethed and foamed" as "great creatures rose from the deep into the air . . . and then fell heavily into the sea."[135] In September, 1884, marine

---

[134]Reeves and Mitchell, "American Pelagic Whaling for Right Whales," 228.

[135]Dolphins were abundant at Hatteras Inlet in the winter of 1874-1875, according to Nathaniel H. Bishop, *Voyage of the Paper Canoe* (Edinburgh: David Douglas, 1878), 189-191.

mammologist Frederick W. True visited the dolphin fishery at Hatteras and found "scores of skulls and fragmentary skeletons" around the station, which was on the ocean side. The season began each year in November or December and usually ended by May, with four to six pilot-boats sweeping the shore waters with 100- to 200-yard segments of eighteen-inch nets. The oil from the captured dolphins was sold at Norfolk or Elizabeth City.[136] George L. Sparks of Philadelphia purchased a contract for all the dolphins captured over the ensuing five-year period, and he intended to sell meat products and oil from the catch, which totaled 600 dolphins by February, 1885.[137] From 1885 to 1891, John W. Rollinson actively captured dolphins at Hatteras and served as superintendent of Colonel Jonathan Wainwright's "porpoise factory" at Creed's Hill, between Hatteras and Frisco.[138] A similar dolphin factory, owned by a Mr. Gardiner of New Jersey, was active at Diamond City for a few years, and in the 1880s and 1890s a porpoise camp was active at Rice Path, about one mile west of present-day Salter Path on Bogue Banks.[139] By 1913 the Hatteras fishery was owned by Joseph K. Nye of New Bedford, under whose management as many as 1,550 dolphins had been taken in a single season. The dolphin and whale fisheries were usually pursued by different individuals, with little attempt to combine the two activities.[140] After several hundred years of intermittent activity, the Outer Banks dolphin industry apparently ceased in 1929.[141]

A more colorful view of local shore whaling at this time was recorded by residents of Shackleford Banks in their traditions concerning the capture of certain right whales. It had become a habit to give particular names to many of the whales, especially the larger ones. One such whale was named the "Little Children Whale" because the boats that chased and killed it were manned mostly by young boys, as the older members of the whaling crews were preoccupied elsewhere when the whale was spotted. Other whales were dubbed the "Lee Whale," the "Tom Martin Whale," the "Big Sunday" whale, the "John Rose Whale," and the "Mullet Pond" whale. The "George Washington Whale" was christened by virtue of being captured on the president's birthday. According to H. H. Brimley, director of the North Carolina Museum of Natural History (later Sciences), the "Cold Sunday" whale was so named because "the day

---

[136]Frederick W. True, "The Porpoise Fishery of Hatteras, N.C.," *Bulletin of the United States Fish Commission*, V (1885), 3-6.

[137]George L. Sparks, "Porpoise Products," *Bulletin of the United States Fish Commission*, V (1885), 415-416.

[138]Rollinson Notebook (microfilm), 28, 60-69, John W. Rollinson Papers, Southern Historical Collection.

[139]Stick, *Outer Banks*, 187; Kay Hart Roberts Stephens, *Judgment Land: The Story of Salter Path* (Havelock: The Print Shop, 1984), 33-36.

[140]Charles Haskins Townsend, "The Porpoise in Captivity," *Zoologica*, I (June, 1914), 289-299; Roy Chapman Andrews, *Whale Hunting with Gun and Camera* (New York: D. Appleton and Company, 1916), 278-283, hereinafter cited as Andrews, *Whale Hunting*; Earll, "North Carolina Fisheries," 490-491.

[141]Mead, "Net Fisheries for *Tursiops truncatus*," 1157.

The best-known of the North Carolina whales was the "Mayflower," named according to custom by the residents of Shackleford Banks. Josephus Willis (above) and his "Red Oar Crew" battled and finally landed the fifty-foot beast near Shackleford in 1874. It was probably the most vigorous fighter ever killed in the area. The "Mayflower" skeleton, now in the North Carolina Museum of Natural Sciences, appears on the adjacent page. Photograph of Willis from *The Heritage of Carteret County, North Carolina* (Beaufort: Carteret County Historical Research Association, 1982), 497. Photograph of skeleton from North Carolina Museum of Natural Sciences, Raleigh.

it was killed was so cold that flying ducks froze solid while in flight."[142] This custom of naming individual whales appears to have been confined largely to the Carolina shore whalers and probably reflected in part the fact that only a few whales were taken in a typical season, thus conveying some uniqueness to each capture.

---

[142]Stick, *Outer Banks*, 190; Odum, *H. H. Brimley Writings*, 110-111; *News and Observer*, November 4, 1956; *Beaufort News-Times*, June 5, 1980. Herbert Hutchinson Brimley (1861-1946), zoologist and director of the North Carolina Museum of Natural History (later, Natural Sciences), was a native of England who arrived in Raleigh in 1880 and resided there until his death. Brimley's publications and interests were broad ranging, and he and his brother Clement Samuel Brimley (1863-1946) rose to prominence as the leading naturalists of the Southeast during their lifetimes. Powell, *DNCB*, I, 227-228; John E. Cooper, "The Brothers Brimley: North Carolina Naturalists," *Brimleyana*, I (March, 1979), 1-14.

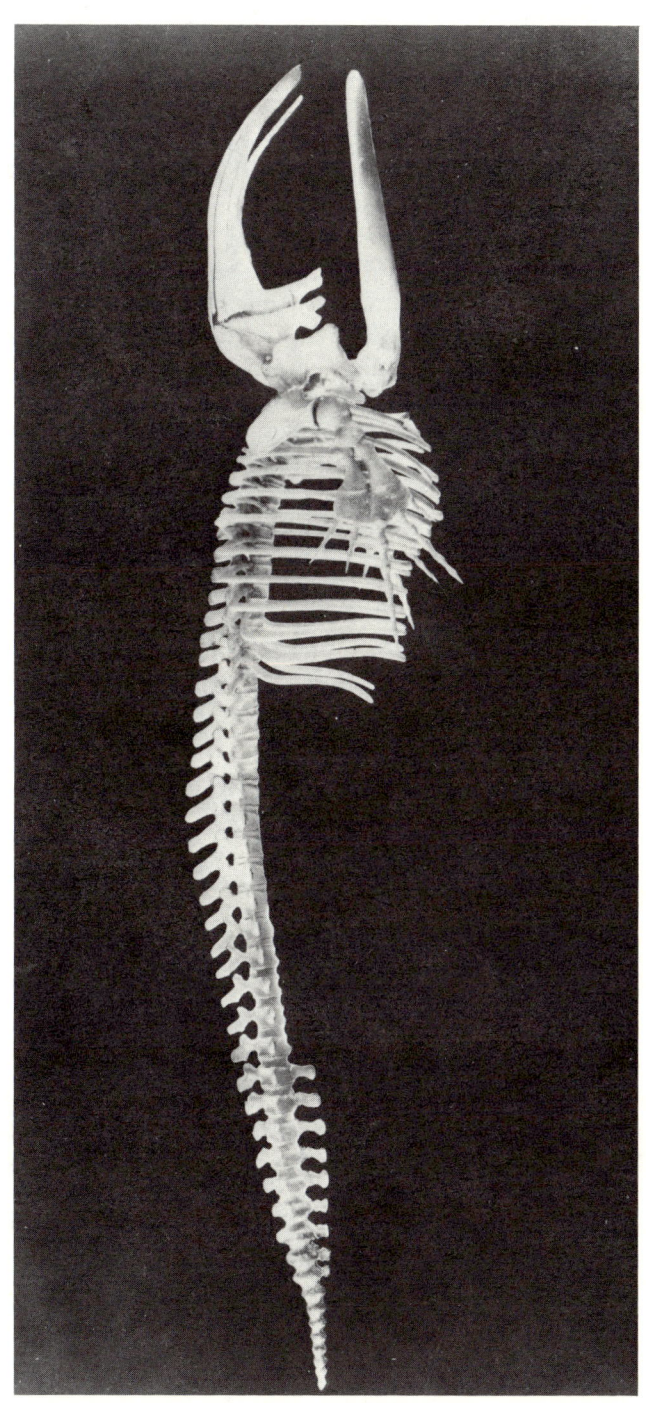

Perhaps the best known of the named whales was the "Mayflower," a large right whale taken near Shackleford Banks on May 4, 1874. Recollections of the incident vary somewhat in detail, but all accounts claimed that this whale was probably the most vigorous fighter ever killed in the area. In the early morning hours, Absalom Guthrie, a veteran whaler, spotted a large herd of whales to the southwest. From his lookout position in the high sand dunes, Guthrie estimated that at least ten female or cow whales were in the group. He sounded the alarm, and five or six boats hit the surf in pursuit. Each boat was manned by a crew of six, four men rowing, with the captain directing from the bow and the steersman in the stern maneuvering the boat. The captains on the hunt of May 4 were Elzie Guthrie, W. O. Guthrie, James Lewis, Josephus Willis, Reuben Willis, and Samuel Windsor, a black fisherman.

The Willises' boats soon caught up with the whales near the outer fringes of the hook or bight at Cape Lookout. Although Reuben Willis was probably the first to hit the whale with harpoon, his boat almost immediately capsized in the ensuing melee. Josephus Willis managed, however, to strike the whale and keep his crew and boat intact during the "Nantucket sleigh ride" that dragged them some six to eight miles out to sea. There the "Mayflower" made its stand, dying after a six-hour battle with whale guns and harpoons. "He fought and slashed about at the six boats for hours, they laying off to one side, and whenever a captain in either boat could pick a chance when his tail was not slashing in his direction, he would rush in and harpoon him."[143]

The whalers credited Josephus Willis with the catch, and his "Red Oar Crew" labored into the night towing the dead giant back to the shore, where large fires were ablaze to guide the weary seamen toward the beach. The crew actually consisted of Willis's five sons, they being the only "all-whaling family" on the coast.[144] Other crews took two more whales the same day, and by the morning of May 5 all had been given names. The "Lady Hayes" yielded up 35 barrels of oil and 650 pounds of whalebone, and the curiously dubbed "Haint Bin Named Yit" produced 25 barrels and 450 pounds of baleen. As the big catch of the day, the "Mayflower" produced 40 barrels of oil and 700 pounds of whalebone.[145]

The "Mayflower" has remained the best known of North Carolina's whales, for its skeleton has been displayed since the 1880s at the North Carolina Museum of Natural Sciences in Raleigh. The remains of the "Mayflower" were presented to the museum sometime between 1874 and 1880 by Colonel John D. Whitfield, president of the Atlantic and North Carolina Railroad, who had instructed the whalers to save and clean the bones. When the North Carolina Department of Agriculture hired H. H. Brimley as a "fertilizer inspector" in March, 1894, his "singular task"

[143]Odum, *H. H. Brimley Writings*, 111; *News and Observer*, November 4, 1956; *Beaufort News-Times*, June 5, 1980.

[144]*News and Observer*, November 4, 1956.

[145]Odum, *H. H. Brimley Writings*, 110-111; *News and Observer*, November 4, 1956.

was to clean and prepare the "Mayflower" bones for proper exhibition at the museum, where the reassembled skeleton continues to be a major attraction.[146]

Perhaps the most detailed account of coastal whaling was provided by Brimley, who visited Shackleford Banks in the spring of 1894 to examine a specimen captured near Wade Shore, a settlement about halfway between Beaufort Inlet and Cape Lookout along Shackleford Banks. Upon his arrival there, Brimley discovered that six whales had been seen and chased that season, but only two whales had been taken. The first was 30 feet in length but yielded up 25 barrels of oil and more than 150 pounds of whalebone. The second whale, however, was enormous—one of the largest right whales on record. The creature was about 53 feet in length, 45 feet in maximum circumference, and weighed an estimated 120 tons. It yielded 39 barrels of oil and 864 pounds of whalebone, all totaling about $1,900 value to the captors.[147] Brimley recounted the story of the whale's capture and death in some detail:

Before sunrise on the morning of March 20, 1894, some of the crew of one of the whale boats belonging to this part of the beach sighted three whales less than a mile from shore. They consisted of a bull, cow, and yearling. In an incredibly short space of time the news was conveyed to the members of the crews of the four boats lying up on the beach, and the men hurried down, ran their boats into the water, jumped aboard and put out in chase. The first boat off was in command of Joe Lewis, and about sunrise this boat worked carefully up to the nearest whale, the cow, until close enough for a harpoon to be thrown. The whale was seen to be a white-bellied one of the largest size, and every precaution was taken to get the first harpoon in deep and fast. The monster rose to blow, the boat worked quietly and carefully up until within a few feet of its shining black body; Mart Guthrie lifted the harpoon and with a powerful swing of his strong right arm he threw the iron in. The boat backed off; up went the flukes of the monster's powerful tail and down she went, the harpoon in her body, while behind it trailed the seven fathoms of one-inch line with a block of wood attached to the end as a drag.[148]

The boat was carefully maneuvered to a position close to where the whale was expected to rise. After remaining submerged about thirty minutes, the wounded whale finally surfaced nearby.

Joe Lewis was now in the bow with the whale-gun and as soon as the boat could be run in close enough he raised the heavy gun and fired one of the long, bolt-like bombs into her body. Again she went down while the gun was reloaded and the boat kept away after her. And so the fight kept on, the whale all the time working out seaward, while the boat kept in full chase, shooting at every available

[146]Grace John, "History of the North Carolina State Museum" (an unpublished manuscript, North Carolina Museum of Natural Sciences, n.d., Raleigh).

[147]Odum, *H. H. Brimley Writings*, 112-113; True, "Whalebone Whales," 246-268.

[148]H. H. Brimley, "Whale Fishing in North Carolina," *Bulletin of the North Carolina Department of Agriculture*, XIV (April, 1894), 4, hereinafter cited as Brimley, "Whale Fishing."

In 1894 the North Carolina Department of Agriculture hired taxidermist and naturalist Herbert H. Brimley to prepare the bones of the "Mayflower" for display in its new museum of natural history. Brimley also visited Shackleford Banks in that year and produced a detailed account of coastal whaling. He served as either curator or director of the museum until his death in 1946. Photograph of Brimley at his desk in 1930 from the files of the Division of Archives and History.

opportunity. Meanwhile the bull and yearling had got a whiff of blood from the wounded cow and they headed out to sea, fighting-mad, and thrashing angrily around. The other three Wade Shore boats kept off after them, but their anxiety to escape was so great that the powerful, five-oared boats, although pulled by well-trained rowers, lost ground all the time, and finally gave up the chase without getting a single opportunity to throw a harpoon or shoot a bomb.[149]

Either bad aim or defective ammunition plagued the whalers, for after they had shot five bombs into the creature, it showed no signs of slackening its seaward pace. Captain Lewis's boat was soon joined by another boat from Morehead City, the latter captained by Mart Willis. The two combined forces to subdue the great whale:

The shore line was getting fainter and fainter and the whale getting weaker from loss of blood. She was keeping under about half an hour at a time, and nearly every time she rose to blow one or the other of the boats made its mark. Finally, at about noon, the end came. She rose, with both boats close up. Joe Lewis fired another bomb into her body, laid down the gun and picking up a harpoon drove the iron home. At the same time Mart Willis fired a bomb, and another harpoon

---

[149]Brimley, "Whale Fishing," 4.

and a lance were likewise put in. Then the boats, which were right up on the whale's back, backed off and lay on their oars waiting for the end. It came soon. She was spouting thick streams of blood, showing that she was badly wounded in the lungs, the vital spot, and after a few flurries with her powerful tail she lay dead on the water, ten miles out to sea, with nothing to indicate the direction of the land but the thin dark line of the tops of the cedars on the banks to the northward and the tapering tower of the Cape Lookout lighthouse to the northeast.[150]

The crews struggled to tow the beast to shore, where she was beached and the blubber and whalebone removed. Brimley's text and the accompanying illustrations provide a good description of this odiferous process, known as "trying out," whereby the blubber was boiled to extract the whale oil.

After a whale is caught then comes the very necessary and more unpleasant operation of cutting up and trying out. The carcass is hauled up as high as possible on the beach, lines fastened to anchors sunk in the sand are made fast and the tide allowed to wash it up as high as it will. The head is cut off and the whalebone cut out of the upper jaws in blocks and piled up like a shock of corn. The tongue is next cut out in pieces, being too large to handle whole. That belonging to the one examined must have weighed about two tons. The tools used in cutting up are known as spades. They are long and broad-bladed chisels, ground very sharp and fitted with a long wooden handle. The whole tool is some six or eight feet long, and the blade six or eight inches across. The blubber is cut in long strips with a pushing, jabbing motion of the spade and then crosswise so as to get it off in square blocks small enough for two men to handle. A hole is cut near one edge, a pole run through it and it is then carried across to the try-kettles, which, in this case, were about one-fourth mile away.[151]

Most whalers attempted to be rather thorough when harvesting fat from a whale. Besides stripping the blubber from the body, they also removed the tongue, lips, flukes, and deep body fat as well. The workers processing the whale described by Brimley finally had to spade out the carcass and drag it a short distance with hooks in order to get at the "under blubber" along the belly. They then processed all this material for whale oil. According to Brimley,

The try kettles are large iron pots of about fifty gallons capacity, and in this case they had two in use set in brick-work over one fire. The blubber, as it is cut from the carcass, is piled up near the try-kettles. It is then "minced," either with a spade in a tub or on a bench with an old scythe blade, and is then thrown into the kettles. As the boiling is finished the oil is dipped out with a long-handled copper ladle and poured into the strainer, which consists of a wide-flaring trough with holes in the bottom, the holes being plugged loosely with bullrushes. The strained oil runs into a long dug-out trough with a partition across the center, the partition

---

[150]Brimley, "Whale Fishing," 5.
[151]Brimley, "Whale Fishing," 6.

also having auger holes plugged with bullrushes. This secondary straining renders the oil perfectly clear, and from the lower end of the big trough it runs through a hole in the side into a small movable trough which connects with the bung-hole of the barrel. The barrel lies on its side in a hole in the ground and as soon as filled is lifted out and replaced by another. The crackling is dumped from the strainer in a pile and used as required, in conjunction with red cedar wood (the common growth on the banks), in keeping up the fire under the pots. On the leeward side of the kettles the steam from the boiling oil, combined with the thick smoke of the burning crackling, makes the smell one to be remembered.[152]

The islanders usually joked that the odor "smelled like money" to them,[153] and the proceeds of each catch were divided according to rules that allowed various shares to everyone involved in taking a particular whale. According to Brimley, each gunner got two shares; one share went to the boat owner; the owner of each full set of tackle got two-thirds share; and the harpooner and steersman received the other third. Every crew member actually participating in the kill received a share, and the owner of each kettle used in trying out was given five gallons of the oil. The whalebone and oil were sold at auction to the highest bidder, and the profits were then distributed. At the time of Brimley's 1894 visit, there were five whaleboat crews active at Shackleford, "one or more" at Cape Lookout, and one at Morehead City.[154]

The whaleboats were constructed locally and designed specifically for whaling. Shackleford residents such as Devine S. Guthrie crafted these sturdy lapstrake pilot-boats from local timbers, often three-fourths-inch juniper with ribs of stripped cedar roots. Most were twenty to twenty-five feet long, double-ended, with high-pointed bows and sterns, and built to carry crews of six men. The design made the boats quite light and buoyant but incapable of rough use. Four men handled the oars, another served as steersman, while the gunner or harpooner worked the bow oar until the craft closed in for the kill.[155] Although there were many variations in style and construction of whaleboats, the double-ended, lapstrake design was a basic form used throughout much of the history of the whaling industry and dated back at least to the mid-1300s, when it was employed by Basque shore whalers.[156]

Whale guns were reportedly introduced on the North Carolina coast sometime after the Civil War, although the exact date and circumstances are not clear. At least two such guns that were used by the Shackleford shore whalers have survived to the present day. Both are single-barreled, muzzle-loading shoulder pieces that fired explosive "bomb lances," which

---

[152]Brimley, "Whale Fishing," 6-7.

[153]*News and Observer*, November 4, 1956.

[154]Brimley, "Whale Fishing," 4-8; Odum, *H. H. Brimley Writings*, 113.

[155]Brimley, "Whale Fishing," 5; Stick, *Outer Banks*, 190; *News and Observer*, August 31, 1969; *Beaufort News-Times*, June 5, 1980.

[156]Willits D. Ansel, *The Whaleboat* (Mystic, Connecticut: Mystic Seaport Museum, 1985), 8, hereinafter cited as Ansel, *Whaleboat.*

Shackleford boatbuilder Devine S. Guthrie is pictured here with one of the sturdy, double-ended, lapstrake boats used in North Carolina whaling. Guthrie and other builders crafted the boats, which carried a crew of six, from local timbers, often juniper with ribs of stripped cedar roots. The basic design dated back to at least the fourteenth century, when it was used by Basque shore whalers. Photograph from North Carolina Maritime Museum, Beaufort. Reproduced courtesy of David Brooks, Morehead City.

were projectiles about 1 inch in diameter and 16 to 18 inches long. Designed with rubber "feathers" to guide the missle to its victim, these bombs exploded deep within the whale's body, thus ending the hunt quickly and often at a safer distance than that required by use of the lance. The gun reportedly used by Josephus Willis in taking the "Mayflower" whale was made by Julius Grudchos and Selmar Eggers, gun manufacturers of New Bedford, Massachusetts. This shoulder gun weighed sixteen and a half pounds and measured thirty-eight inches in length. Its sturdy hardwood stock, trimmed in heavy brass plate, supported a cast-steel rifled barrel of one-inch caliber. Another firearm, developed by Oliver Allen of Norwich, Connecticut, was popular with whalers on the Carolina coast. That model was called a "Brand Whaling-Gun" after 1852, when Christopher C. Brand of Ledyard, Connecticut, received ,a patent for an improved bomb lance to be used in the weapon. Other types of guns also may have been purchased by North Carolina shore whalers.

One report of a seventy-five-pound weapon, for example, suggests a model designed for mounting to a boat.[157]

Two types of harpoons or "irons" were in common use: the "two-fluted" or older style of fixed-blade spear and the more advanced "toggle iron," which had a hinged headpiece that became firmly anchored when driven into the whale by the harpooner. A short line or warp of cord, usually less than forty fathoms in length, was tied to the harpoons, a method that differed from that employed by the New England whalers, who used very long lines to maintain a constant connection between the whale and the boat. A square block of wood, called the "drag" or "drogue," was attached to the end of the warp. The drag block could thus be retained under the head cap of the boat until it became "advisable" to release the harpooned whale, as when the creature sounded or dived. The whalers then simply threw the drag overboard, where it served the functions of slightly impeding the whale's movement and of marking the beast's location like a buoy. Similar techniques apparently date back to the early days of organized whaling, as Basque shore whalers may have used wooden drags and short warps in the 1300s. The final piece of equipment was the lance, a long-shanked steel blade that the crew used for the kill by spearing the animal's lungs. In 1898 North Carolina whalers employed such equipment and mode of operation to capture another whale that Brimley also investigated and described.[158]

On February 14, 1898, a female right whale was sighted near Cape Lookout and pursued by Captain Tyree Moore's "Red Oar Crew" and a boat captained by John Lewis, who used a whale gun and harpoons to dispatch the animal. The crews towed the whale to shore and beached her near the western end of Shackleford Banks close to a small brackish lake known as the "Mullet Pond," which was thus deemed suitable as a name for the whale.[159] The "Mullet Pond" whale measured about forty-six feet long and provided some 650 pounds of whalebone and twenty-seven barrels of oil.[160] When Brimley returned to Raleigh after investigating the specimen, he decided that it might be a "ripe and fruity personal gamble to have the bones of the skeleton roughed out with the idea of selling them to some museum in need of a specimen." Brimley convinced his brother C. S. Brimley to join in the venture, and together they acquired

---

[157]Brimley, "Whale Fishing," 5-6; Stick, *Outer Banks*, 190-191; *Beaufort News-Times*, June 5, 1980; Davis, Hamilton, and Pitts, *Carteret County*, II, 417; Thomas G. Lytle, *Harpoons and Other Whalecraft* (New Bedford: Old Dartmouth Historical Society, 1984), 77-119.

[158]*News and Observer*, September 21, 1941; Davis, Hamilton, and Pitts, *Carteret County*, II, 420; Ansel, *Whaleboat*, 8; Brimley, "Whale Fishing," 5-6; *Beaufort News-Times*, June 5, 1980.

[159]Will Thomson, "A Whale for Iowa," *Palimpsest*, 68 (Summer, 1987), 50-59, hereinafter cited as Thomson, "A Whale for Iowa"; *News and Observer*, February 15, 1898.

[160]True, "Whalebone Whales," 246, 248; "Skeleton of Balaena Biscayensis, killed on the coast of North Carolina, near Cape Lookout, February 15, 1898," advertisement circular by H. H. and C. S. Brimley, Raleigh, Brimley Papers, hereinafter cited as Brimley and Brimley, "Skeleton of Balaena Biscayensis"; *News and Observer*, February 15, 1898.

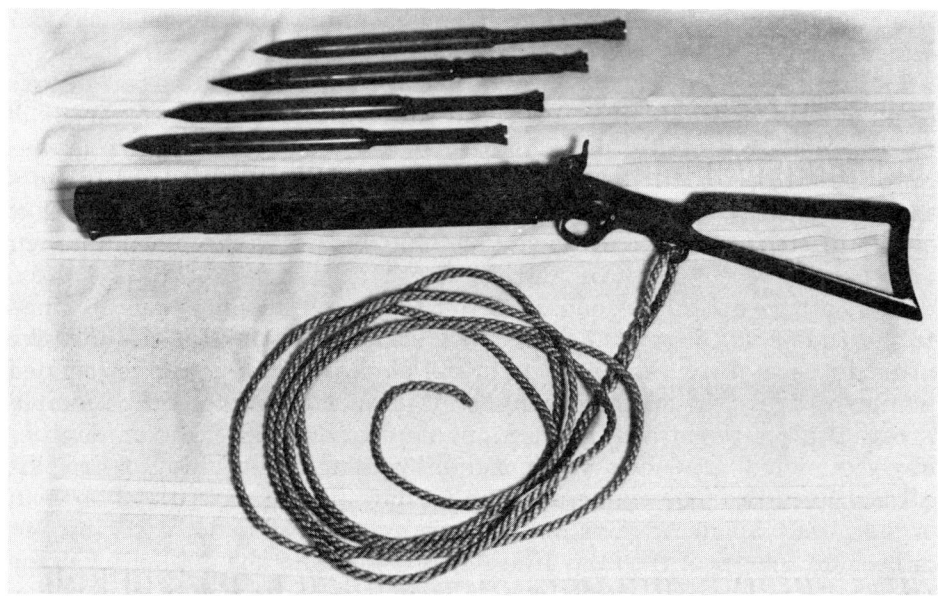

Whale guns reportedly were introduced on the North Carolina coast after the Civil War. Such shoulder weapons fired explosive "bomb lances" that detonated deep within the whale's body. This gun, now in the North Carolina Museum of Natural Sciences, was called a "Brand Whaling-Gun" after 1852 and was a popular model among whalers. Photograph from North Carolina Museum of Natural Sciences.

from one J. H. Potter of Beaufort the rights to the remains, "one of the most smelly collections of half-cleaned bones" imaginable.[161]

Problems arose, however, when Brimley tried to have the skeleton shipped to Raleigh, for the railroad wanted the material classified as "whalebone," which would have incurred prohibitive charges for shipping. After a rather drawn-out correspondence, Brimley managed to persuade the railroad to handle the bones under the classification of "fertilizer," a description that was in keeping with the associated odor and that permitted shipping at a much lower rate. After circumventing further problems with storing the bones in Raleigh, the Brimleys advertised the skeleton for sale at $250. By May, the bones were purchased by Charles Nutting of the Museum of Natural History at the University of Iowa in Iowa City, where the reassembled skeleton of the "Mullet Pond" has been displayed since 1911.[162]

According to William Hansen, a Mormon missionary in the Beaufort-Harker's Island area, another huge whale was captured only four days later, when on Friday, February 18, 1898, Captain Thomas Lewis discovered the creature "close to shore." The crew under Captain Lewis's

[161]Odum, *H. H. Brimley Writings*, 107; Brimley and Brimley, "Skeleton of Balaena Biscayensis."

[162]Thomson, "A Whale for Iowa"; *H. H. Brimley Writings*, 107.

command soon killed the whale and pulled it aground at "The Banks," where Hansen noted that the carcass measured about sixty feet in length and fifteen feet in height. The whalers anticipated a yield of 4,800 gallons of oil and considerable whalebone, to a total value of $1,800. The pieces of baleen measured eleven feet by three feet. Hansen apparently climbed onto the whale's back and noted that "it seemed as though we were on a small Island." Curiously, Brimley and other scientists did not know of this particular capture; at least they made no reference to it in their writings.[163]

Most of these whaling activities were centered on Shackleford Banks from at least the time of the Civil War until the demise of the industry after the turn of the century. Whaling, fishing, crabbing, and associated ventures were of sufficient magnitude that Shackleford gradually acquired a rather large population of permanent residents. Several communities evolved, including Wade's Shore, a small settlement about four miles east of Beaufort Inlet. The major community, however, was known as Lookout Woods until about 1885, when it was renamed Diamond City by Joe Etheridge, keeper of the local lifesaving station, for the diamond pattern on the Cape Lookout lighthouse.[164] Diamond City was located at the eastern end of Shackleford Banks, very close to the lighthouse. A typical banks community, Diamond City's population numbered in the hundreds, perhaps as many as 500 during its heyday; and the city boasted homes, family graveyards, stores, factories, and a school. The city eventually extended from Barden Inlet westward across almost half the length of Shackleford Banks, and the size and antiquity of the community can be gauged by the fact that local tradition claims that one of the large graveyards had as many as 500 interred bodies.[165]

By the end of the nineteenth century, Diamond City was on the wane, in part from the decline of the whale fishery and perhaps in equal measure because of the weather. In the 1890s a number of severe storms struck the region, flooding gardens and inflicting considerable property damage. People began leaving Shackleford, sometimes actually removing their homes and transporting them across the sound to resettlement sites at Harker's Island, Morehead City, and Salter Path on Bogue Banks. A particularly bad hurricane hit the area in August, 1899, and the damage was so severe that the populace began to abandon the area in earnest. Over the next few years, virtually all the remaining residents moved to the mainland or more sheltered island locales, leaving Shackleford to the elements, the once bustling Diamond City to sand and sea oats.[166]

Whalers apparently continued to visit the island in the spring months, however, setting up their camps and lookouts just as many previous

---

[163]William Hansen, "Missionary Journal, August 1897 to Dec 1898," 70, a manuscript journal in possession of Mickey Hansen, St. Anthony, Idaho. Quoted with permission of Joel Hancock.

[164]Stick, *Outer Banks*, 184-194; *News and Observer*, November 4, 1956; August 31, 1969; Holland, *Survey History of Cape Lookout*, 18-19.

[165]Stick, *Outer Banks*, 187-189; Holland, *Survey History of Cape Lookout*, 18-19.

[166]Stick, *Outer Banks*, 192-194.

Although old-time whale hunting ended in North Carolina waters in 1925, whales occasionally continued to be stranded or washed ashore along the coast. This dead sperm whale appeared on Wrightsville Beach in 1928. Photograph from the files of the Division of Archives and History.

generations had done. Although some residents claimed that whales were just as likely to be seen as in previous decades, the North Atlantic whale populations by the early twentieth century had been so extensively over-hunted that little profit was to be had, and in many years no whales were observed. Perhaps more significant was the collapse of the market for whalebone, when in 1907 a change in women's fashion virtually eliminated any need for baleen corset stays.[167]

Among the few captures during the sporadic final years of the shore industry was the "Tom Martin" whale, a forty-three-foot right whale taken in April, 1908. After a ten-day trying out process it yielded forty-five barrels of oil.[168] John Rose captured a thirty-nine-foot male right whale inside the bight at Cape Lookout on May 17, 1908, but the blubber was so thin that no effort was made to process the oil.[169] The last whale reportedly captured on the North Carolina coast was killed on March 16, 1916, when

[167]In 1907 a young Parisian couturier, Paul Poiret, introduced a new slim look to his fashion designs, quickly leading to the disappearance of the whalebone corsets that had been required for the narrow-waisted, S-shaped fashions of late Victorian and Edwardian style. Elizabeth Ewing, *Dress and Undress: A History of Women's Underwear* (London: Bibliophile, 1981), 113; John R. Bockstoce, *Whales, Ice, and Men* (Seattle: University of Washington Press, 1986), 335-336; Stick, *Outer Banks*, 194.

[168]Davis, Hamilton, and Pitts, *Carteret County*, II, 419.

[169]H. H. Brimley, "A Right Whale on the North Carolina Coast" (an unpublished manuscript, North Carolina Museum of Natural Sciences, Raleigh, 1908).

Charlie and John Rose took a fifty-seven-foot right whale in the shallows at Cape Lookout. The yield of thirty-eight barrels of oil was apparently the last such oil procured by active shore-based whale fishing on the Carolina coast.[170] The last shore crew for whaling at Cape Lookout was disbanded when a fire destroyed most of its gear in 1917.[171]

Shortly after the demise of shore whaling on the North Carolina coast, the sailing period of American deepwater whaling also ended. After the turn of the century, a few sailing ships from New England continued to work the Hatteras ground off the northeast coast of North Carolina during the spring and summer months.[172] Perhaps most notable of these was the *John R. Manta*, a ninety-eight-ton schooner of New Bedford registry and captained on most of her voyages by Antonio J. Mandly, who was also the owner. The majority of the voyages over the vessel's two decades of active whaling were to the Hatteras and Western grounds in search of sperm whales. The *John R. Manta* made its final whaling voyage to the Hatteras ground in the summer of 1925. During this last expedition, the *John R. Manta* cruised many miles from shore and principally in portions of the Hatteras ground lying off the coast of Virginia. Nevertheless, its capture of sperm whales in that area in 1925 may reasonably be considered the closing chapter of old-time whale hunting in Carolina waters.[173]

In retrospect, one of the most intriguing aspects of North Carolina shore whaling is the question of why the industry survived for over two and a half centuries, despite the physical hazards,[174] expense, erratic results, and rather meager profits. North Carolina shore whaling was apparently never more than a minor activity, mostly to provide supplemental income to fishermen during a slack period in their regular occupation. Compared to the complex business ventures of New England whaling, the industry in North Carolina was rather insignificant as an economic phenomenon. Perhaps one factor contributing to this persistence was suggested by Brimley, who stated, "I have always believed that the thrills and excitement accompanying the chase and capture of these monsters of the deep had a great deal to do with the regularity with which those hardy coastdwellers made ready for the chase year after year as the days of early spring rolled around."[175]

With the decline of commercial whaling and the growth of conservation, some whale species are showing evidence of recovery from centuries of

---

[170]Davis, Hamilton, and Pitts, *Carteret County*, II, 205, 420.

[171]David, Hamilton, and Pitts, *Carteret County*, I, 37.

[172]Andrews, *Whale Hunting*, 239.

[173]William Henry Tripp, *There Goes Flukes* (New Bedford: Reynolds Printing, 1938), 1-261; Downey and Adams, *Whaling Logbooks*, 198-199, 438-439. Logbooks of the *John R. Manta* are in four locations: Kendall Whaling Museum, Sharon, Massachusetts; Old Dartmouth Historical Society, New Bedford, Massachusetts; Smithsonian Institution Archives, Washington, D.C.; Providence Public Library, Providence, Rhode Island.

[174]Despite the commonly recognized hazards of whaling, the local tradition in the Cape Lookout area is that "not a man was lost" during the history of shore whaling in the area. Davis, Hamilton, and Pitts, *Carteret County*, II, 418.

[175]Odum, *H. H. Brimley Writings*, 109.

exploitation. Although the status of the right whale remains precarious, numerous records in the North and South Atlantic suggest that the species may yet escape extinction. In recent years right whales have been observed along the North Carolina coast, as they pass from winter calving grounds off Georgia and Florida on the way to their summer resort in waters off the northeast coast of the United States and Canada. Now these rare and intriguing creatures may occasionally be seen plying their way north in spring past the Carolina sand banks where cries of "Thar she blows" once rang out.[176]

----

[176]Female right whales appear to have been the major documented victims of the North Carolina shore whalers, who often relied on the maternal instinct of the cow to stay behind with her calf when the pair came under attack. Whalers sometimes would harpoon the calf but not kill it, using the hapless young whale as "bait" for its mother. Such selective harvesting probably further exacerbated the decline of the species. Davis, Hamilton, and Pitts, *Carteret County*, II, 418-420; Brimley Papers; *Watauga Democrat* (Boone), February 20, 1985; Connie Mason to Marcus B. Simpson, Jr., and Sallie W. Simpson, July 19, 1987, letter in possession of the authors; Reeves and Mitchell, "American Pelagic Whaling for Right Whales," 221-222.